Final Report

on

Determining the Yield Strength of
In-Service Pipe

1.0 INTRODUCTION

The Gas Pipeline Safety Research Committee (GPSRC) is one of several committees established by the ASME Center for Research and Technology Development. The scope and functions of the Research Committee are to

- Conceive, plan, and sponsor research required to increase the state of knowledge and practice related to pipeline safety
- Promote technology transfer
- Promote research concerning gas pipeline safety issues identified by other organizations
- Maintain liaison and cooperate with other relevant organizations regarding gas pipeline safety.

The GPSRC identified the need for a project to determine the yield stress of in-service transmission pipelines. The committee's objective was to determine if hardness testing on in-service line pipe could be an acceptable alternative to the destructive testing or low yield stress requirements of DOT/OPS. Once this objective is attained, the results of the project could be recommended for use in support of a proposal to DOT/OPS for a rule change and to provide applicable enhancement for confirmation of yield stress in the gas piping code.

The GPSRC approved the project and formed a steering committee to manage the project. The ASME Center for Research and Technology Development issued an RFP and subsequently solicited funding support. The steering committee selected Battelle as the Principal Investigator.

1.1 Background

The Department of Transportation's Office of Pipeline Safety (DOT/OPS) has developed requirements for determining the tensile properties of pipe (49 CFR 192, Appendix B) when the actual properties are not known or are not properly documented. These requirements dictate extensive sampling. For example, on pipelines containing over 100 pipe lengths, the requirements specify that one set of tensile tests be conducted for each 10 lengths of pipe. Additional samples are required for pipelines with fewer pipe lengths. The technical details of the tensile testing for each sampled pipe length are set forth in API Specification 5L. Obtaining samples for such testing is both destructive and expensive in that excavation and removal of tensile coupons require either hot-tapping to remove sub-sized specimens, or that the pipeline be blown down, samples cut, and repairs made. Thereafter the dig sites must be closed, graded, and returned to their initial state.

As directed in 49 CFR 192, Appendix B, the estimated yield stress for the pipeline is the smaller of two statistical test values calculated from the resulting tensile property data:

1. Eighty percent of the average yield stress across the sampled pipe lengths, and
2. The minimum yield stress among the sampled pipe lengths.

If a calculated yield-to-tensile-stress ratio equals or exceeds 0.85 for any of the pipe lengths sampled, the entire pipeline is subject to extensive use restrictions, including a penalizing maximum stress limit of 6,000 psi . In contrast, the current version of API Specification 5L sets the upper limit on yield-to-tensile-stress ratio at 0.93 for new construction or replacement line pipe.

Without tensile testing, the alternative is to use a yield stress of 24,000 psi. This requirement can be quite conservative, resulting in pressure reduction or other actions that could adversely impact pipeline operations.

1.2   Objective

The objective of this project was to establish a field useable procedure and data evaluation process for nondestructive evaluation of in-service line pipe to estimate the tensile properties by using a field-based hardness measurement as the primary test method. The procedure and process were designed with a view to their eventual acceptance by the DOT/OPS. To that end, the procedure and process developed by this project were to be cast into a draft proposal for rulemaking with scope and details to facilitate its eventual acceptance as an alternative to the current tensile testing procedure. Acceptance of this proposal could be viewed as the *practical objective* of this project.

1.3   Approach and Scope

Meeting the technical objective requires that a relationship be established between measured hardness and the corresponding tensile stress for measurements made on pipe in the field. Once such a relationship is established, a procedure to analyze the data is required, as well as a test practice to consistently implement the procedure. The relationship can be developed following either of two approaches.

In one approach, a field hardness measurement tool could be taken into the field and hardness measured on coupons that are cut out and used to measure the corresponding tensile yield stress. This approach has the advantage of providing direct measurements to develop the desired relationship. Its disadvantages include the fact that the relationship developed is specific to the field tool used, and the fact that its development involves a very significant cost for even a relatively small sample of pipe steels.

In a second approach, the relationships could be developed in two steps. First, hardness and the tensile properties could be related based on laboratory measurements and thereafter the relationship between laboratory and field hardness could be determined. This two-step approach has the advantage of using an extensive database of laboratory hardness and tensile properties, which includes a wide range of pipe steels. It is also more cost-effective. Its disadvantage is the fact that the relationship between laboratory and field hardness must be established. However, this is not a significant issue, because this relationship must be determined in any event and is nondestructive. Given the balance between advantages and disadvantages, Battelle chose to implement the second approach.

This project developed a field measurement procedure and data evaluation process to nondestructively determine the yield stress of steel materials commonly used in pipeline systems. The scope of these materials was intentionally limited to grade X52 or lower, this being the likely range of older pipe for which the properties are unknown. This project did not simply produce a hardness-to-tensile property relationship — rather it provided all of the elements of the procedure and data analysis process needed to nondestructively determine the otherwise unknown yield stress of in situ line pipe.

Only hardness testing methods that are suitable for field use and are commercially available were considered. While more advanced hardness testing methods involving measurement of indentation displacement and load during the test are now becoming available, such methods were considered to be outside the scope of this project.

Limited field studies and related data analyses using the available literature indicated that a correlation between yield stress and hardness was feasible. These same studies, however, indicated that more specific correlation might be required to represent, for example, the various classes of pipe steel and components used in the industry. Such considerations were included in the analyses described below.

Thus, the approach to meet the technical and practical objectives reflects the need to keep the procedure simple when there is little potential gain, as is the case for the lower strength grades. More complex and precise determination of yield stress is possible for the higher strength grades beyond the level targeted by Appendix B of 49 CFR 192 (i.e., at and above X52), but such determinations are beyond the scope of this project. Because the practical objective is to develop a draft proposal for rulemaking for an alternative, nondestructive procedure that potentially could be accepted by the DOT/OPS, a technically defensible relationship cast in a statistical format was developed that permits characterization of lower bounds on the relationships estimated at different confidence levels.

Because no specific field-hardness measurement equipment is recommended, the process of estimating yield stress from field-measured hardness involves two steps, specifically:

1. Determine a lower bound on laboratory-measured hardness using field-measured hardness data and a statistically characterized relationship between laboratory and field hardness; and

2. Determine the lower bound on yield stress based on a statistically characterized relationship between yield stress and laboratory hardness.

As indicated above, field hardness measurements are first translated into a laboratory measure of hardness, which is in turn translated into yield stress. Desired levels of safety, consistent with those assumed by the current testing requirements, are maintained by using statistical methodology in each translation.

The effort required to establish the relationship between field and laboratory hardness likely varies considerably from one type of field hardness measurement tool to another. Some equipment may be more sensitive to operator differences, axial and circumferential variation in hardness, and surface preparation, to name just a few potential factors. At a minimum, paired samples representing laboratory and field measurements on pipe lengths from a broad range of hardness measurement equipment and heats of steel are necessary. With data from such samples, the relationship between field hardness and laboratory hardness can be characterized and lower bounds estimated on the laboratory hardness associated with any given field hardness for each type of field equipment.

Once the relationship between laboratory and field hardness is known, the corresponding yield stress in found from the relationship between laboratory hardness and yield stress.

5

Establishing a relationship between yield stress and laboratory hardness also requires data collected from a broad range of manufacturers and heats. These data should reflect X52 and below grade steel manufactured in the early twentieth century to be as representative as possible of a broad range of line pipe.

Data were gathered from archives identified in a literature search and included data contributed by pipeline companies and manufacturers, typically culled from mill certification sheets. Pooled, these data permitted development of the hardness to yield stress relationship.

Theoretical considerations and published data indicate a relationship between yield stress and ultimate tensile stress could be utilized, if a hardness-to-tensile stress relationship could be characterized. Alternatively, there was the potential to develop a direct hardness-to-yield-strength relationship. The project's approach was to use the collected data to develop both paths to a relationship between hardness and yield stress and then use the path producing the most sound relationship from a technical and statistically valid perspective.

1.4   Outline of Report

The collection and management of the data used to establish the hardness-to-yield-stress relationship are outlined in Section 2. The establishment of a hardness-to-yield-stress translation, including both the indirect path through tensile stress and the direct path, is presented in Section 3. Also discussed in Section 3 is the relationship between yield-to-tensile-stress ratio and hardness. This is necessary because of the current testing requirements. Section 4 reviews the data collection and analysis used to develop a laboratory-to-field-measured hardness relationship using an example technology. Section 5 discusses the sampling methodology recommended for a hardness-based testing standard as well as its motivation, and Section 6 summarizes the recommended procedure.

The terms yield strength and yield stress are used interchangeably throughout this report. Although it is technically correct to use the term yield stress, yield strength is the common term used in the industry and is maintained here.

2.0 DATA COLLECTION AND MANAGEMENT

As outlined in Section 1.3, data were collected to develop the relationship between laboratory hardness and yield stress as well as ultimate tensile stress. Ideally, the collected data would be representative of the line pipe in pipelines that are expected to utilize the testing standard being developed by this project.

2.1 Data Collection

The ideal statistical sampling scheme used to collect the data sought for this project is a stratified sampling plan from among the heats and grades in the in-service pipe. Such an ideal is impractical and prohibitively expensive, both in time and money, and possibly also logistically impossible (because of proprietary data concerns). Accordingly, data were sought from three sources:

1. Archival data at Battelle and pipeline companies, including the sponsors of this project;
2. Data published in available literature; and
3. Data retained by pipe manufacturers.

Permission to utilize published or archived data not in the public domain was secured as required.

2.2 Database Development

As the data were being collected, an electronic database was created to organize the data. Every attempt was made to preserve as much of the available information associated with the properties sought. In addition to tensile properties and hardness measurements, information retained included the measurement procedures, sample location, chemical properties, percent

8

elongation, pipe diameter, wall thickness, certifying standard, type of weld, heat treatment, steel grade, date of manufacture, date of installation, manufacturer, and customer/data source.

The mixed variety of data media, information provided, and sources resulted in a tiered database. Though the majority of the data sources were printed copies of mill certification reports, electronic copies and microfiche archives of mill reports also were collected. Data were collected from printed failure reports, summary reports from earlier studies, and independently compiled readings. For the purposes of the database, each specific source of data, whatever the format or media, was considered a distinct "document". Within each "document", the data could often be grouped by a set of common descriptors (e.g., heat, steel grade, manufacturer, date of manufacture, customer, and certification standard). Those groupings could be further divided into measurements (e.g., tensile properties and chemical properties) collected at a specific location. This tiered database enables distinction of collected information at the "document" level, at the "description" level, and at the "section" level. A given "document" can reference multiple heats ("descriptions") from which multiple measurements ("sections") were collected. The multiple measurements can reflect repeated sampling of a given length at different axial or circumferential locations, as well as sampling of distinct lengths from the same heat.

The final version of the database has 55 data fields. A screenshot of the Microsoft Access® database's main data entry form is presented in Figure 2.1. However, as some of the data collected contained sensitive or proprietary information, a sanitized version of the database was also created. Figure 2.2 presents a sanitized version of the same form portrayed in Figure 2.1. Sanitizing the database entailed removing or making generic any information regarding the manufacturer, customer, or data contributor.

2.3 Data Summary

Within the original data, there are 285 documents. Of these documents, 3.1 percent are classified as failure reports and 83.4 percent as mill reports. The remainder cannot be simply classified. The mill reports represent 2,881 distinct heats of steel. In total, the database contains 4,731 measurements of yield and tensile stress. Thirty-one of those measurements are either longitudinal results or based on a 0.2 percent offset yield definition. These were excluded from further consideration (but retained in the database). Tables 2.1 and 2.2 summarize the remaining data by decade of manufacture and pipe grade, respectively. The information provided was inadequate to isolate cold expanded pipe from the population. Likewise, information from failure reports was inadequate to determine if the pipeline experienced prior hydrostatic testing (either pre-service or in-service) to a specific pressure level expressed in terms of SMYS.

Of the 4,700 measurements involving yield and ultimate stress, 836 also had hardness measurements associated with them. Five different hardness scales were observed in the data. As the majority (92.2 percent) were recorded using Rockwell B, it was decided to convert the other measurements into Rockwell B using the equations detailed in ASTM E140-95, Standard Hardness Conversion Tables for Metals. Table 2.3 lists the number of samples collected from each of the five different hardness scales. Two converted measurements resulted in a Rockwell B hardness over 100 HRB, which were excluded as outlined in the ASTM standard.

After excluding the 0.2 percent offset and longitudinal yield measurements and the two excessive hardness measurements, 4,698 yield and tensile stress measurements remained. Of these, 834 also had Rockwell B hardness readings.

3.0   DATA ANALYSIS

The objective of the data analysis is to develop a procedure for determining the yield stress of a pipe if the hardness is known. Two approaches were investigated to develop this procedure. The first approach established a relationship between the ultimate tensile stress and yield stress. This relationship, used in conjunction with a relationship between hardness and tensile stress, is used to relate hardness to yield stress. This approach has a theoretical underpinning in that there is already an established relationship between hardness and ultimate tensile stress. The second approach directly related hardness to yield stress. This approach was affected less by statistical uncertainty in the available data and was ultimately adopted.

As noted above, empirical relationships among yield stress, ultimate tensile stress, and laboratory-measured hardness need to be developed and characterized. For example, is the relationship between yield and tensile stress linear as is expected? What about the relationship between tensile stress and hardness? Simply characterizing the relationships, however, is not sufficient. Statistical tolerance bounds on the observed relationships must be developed if a viable conversion process is to be established. The specific confidence level and percentile that make up the tolerance bound need to be determined. The values to which these parameters are set will have profound impact on the usefulness of such a procedure.

Analysis of the relationships began by examining scatterplots relating yield, tensile, and hardness. Figures 3.1, 3.2, and 3.3 plot yield versus tensile stress, tensile stress versus hardness, and yield stress versus hardness, respectively, for the data collected by this project (4,698 tensile with yield stress and 834 hardness measurements). In addition, Figure 3.4 plots yield versus tensile stress for only those samples in which hardness was also measured. Each point in Figures 3.1 through 3.4 represents an individual pair of measurements, the majority of which (62 percent) were each collected from an individual heat. In other words, the variability evident in Figures 3.1 through 3.4 reflects heat-to-heat (and manufacturer-to-manufacturer) differences rather than axial or circumferential variations within a particular heat. Distinct plotting symbols and colors are used to indicate the grade of steel considered, including a default symbol for those cases for which the grade is unknown.

Inspection of Figures 3.1 through 3.4 suggests the yield-tensile, tensile-hardness, and yield-hardness relationships are effectively linear in character. Before further exploring these relationships, however, it is necessary to assess whether samples whose yield-to-tensile ratio exceeds 0.93 influence the results. The current DOT 49 CFR 192 Appendix B testing requirements include separate consideration of those pipe lengths with yield-to-tensile ratios above 0.85. As will be discussed further in Section 3.4, however, API 5L now allows a maximum value of 0.93. Figures 3.5 through 3.7 plot yield versus tensile stress, tensile stress versus hardness, and yield stress versus hardness, respectively, with distinct plotting symbols and colors used to identify those samples with a yield-to-tensile ratio greater than or equal to 0.93. There are too few pipe lengths with yield-to-tensile-strength ratios greater than 0.93 to clearly assess whether a distinct relationship is evident in Figures 3.5 through 3.7; nonetheless, all such pipe lengths (14 with only yield and tensile stress measurements; 4 with yield, tensile, and hardness measurements) were dropped from further consideration.

3.1 Tensile-to-Yield Stress Relationship

The plot of yield versus tensile stress (Figure 3.5) shows a split in the cloud of points, suggesting at least two distinct relationships with very different slopes (despite excluding lengths with yield-to-tensile ratios that are at or above 0.93). Assessment of the relationship between yield and tensile stress proceeded both empirically and theoretically. For each factor (whether discernable or theoretical), the four scatterplots (Figures 3.1 through 3.4) were reproduced with distinct plotting symbols identifying levels of the factor under consideration. Relationship plots for hardness-to-tensile and hardness-to-yield were prepared in order to jointly assess whether the factors being evaluated might also explain aspects of these relationships.

The assessment identified two factors as critical to the relationships:

1. Application as heavy-wall plant versus typical transmission line piping; and
2. Decade of manufacture.

12

The collected data included measurements of material used for plant piping, evident in the diameter to wall thickness ratio, certifying standard, and company submitting the data. Given the change in character of steel production through the years, collected samples were categorized according to the decade of their manufacture (see Table 2.1) if available. Figures 3.8 through 3.11 show the importance of application (plant versus transmission line piping), while Figures 3.12 through 3.15 show the relevance of the decade of manufacture. The split evident in the relationship between yield and tensile stress (Figures 3.1 and 3.5) seems primarily due to plant versus transmission line differences (Figure 3.8). The distinction between what we call plant versus transmission piping is reflected in such factors as diameter to wall thickness ratio (d/t). Plant piping tends to have a small d/t as opposed to transmission piping. This results from the difference in processing (seamless versus welded) required to manufacture a pipe with a large d/t versus a small d/t.

Statistical tests were conducted to validate the importance of these factors. In each case, the test assessed whether distinct linear relationships (for each level of the factor under consideration) explained significantly more of the tensile-to-yield data than did a single linear relationship. For example, do distinct intercepts and slopes for plant and transmission piping data represent these data more accurately than did a single intercept and slope to warrant the increased complexity (i.e., an additional intercept and slope)? Table 3.1 presents the intercept, slope, and uncertainty of the distinct relationships estimated for plant versus transmission piping, and for the seven decades of manufacturing (plus unknown) that were considered. Figure 3.16 superimposes on the yield versus tensile stress data the fitted lines for plant and transmission line samples. Also, superimposed on Figure 3.16 is the line fitted to all the data. Distinct line types delineate the three linear relationships. A similar plot for the decade of manufacturing is presented in Figure 3.17. The tests for both factors (application type and decade of manufacture) were statistically significant ($p < 1$ percent). More importantly perhaps, the same test for decade of manufacture using only transmission line data was also statistically significant ($p < 1$ percent).

Having identified two factors impacting the tensile-to-yield stress relationship, the question arises regarding which subsets formed by the factors should be considered. The scope of this project, unfortunately, precludes consideration of all such subsets. For this reason, the plant piping data were set aside. The newest transmission line samples, those from heats manufactured in the 1980s and 1990s, also were not considered further. Pipelines constructed from recently manufactured materials are unlikely to need the test procedure this project is developing. Also, the tensile-to-yield stress relationships for these decades are sufficiently distinct to warrant setting aside these data. All further tensile-to-yield stress analysis, therefore, considered transmission line samples manufactured prior to 1980.

Even after restricting the data to transmission line samples of heats manufactured prior to 1980, a handful of data points remained problematic. A statistical test procedure was conducted to identify potential "outliers." Among the eight points identified by this procedure were five points that had been "questionable" since their input into the database. The paper records from which these points were drawn were difficult to discern or the values were such that the recorded digits were likely transposed. Only these eight "outlier" points were excluded from further analysis. The decision to exclude these observations was unanimously approved by the ASME/GPSRC Steering Committee on February 2, 1999.

The estimated tensile-to-yield stress relationship is portrayed in Figure 3.18 by the solid line superimposed on the data. The estimated parameters of the line are noted in Figure 3.18. An asymptotic lower 99 percent confidence bound on the 0.5 percentile yield stress is also traced (dashed line) for each tensile stress. *Note: The 0.5 percentile corresponds to one occurrence in 200, while the 99 percent confidence bound relates to the fact that 99 cases in 100 can be expected to occur. Similar interpretations apply with correspondingly different values in the following presentation.* For a given ultimate tensile stress, therefore, at most 0.5 percent (with 99 percent confidence) of the lengths of transmission pipe from heats manufactured prior to 1980 have yield stresses below that represented by the dashed line. The data used to estimate this relationship and lower tolerance bound are summarized in Table 3.2.

In deriving the tolerance bound portrayed in Figure 3.18, the scatter about the line was assumed to follow a Normal (or Gaussian) distribution. Figure 3.19 presents a probability plot assessing the viability of this assumption. If the data are inconsistent with a Normal distribution, the plotted points will trend off the superimposed line. The limits of the probability plot represent how well the extremes of the data fit the assumed tails of the Normal distribution (e.g., a value of 2.57 represents the 0.5 percentile). A review of Figure 3.19 suggests the assumption is reasonable, though a Kolmogorov D statistical test rejects ($p < 1$ percent) the assumption.

Table 3.3 reports the lower tolerance bound yield stress for a range of tensile stresses. A more complete table of values is included in Appendix A. Since the confidence level and percentile to be used in any testing standard are still to be determined, Table 3.3 presents results for nine different combinations of confidence level (99.9 percent, 99 percent, and 95 percent) and percentile (0.1 percent, 0.5 percent, and 1 percent). The specific combination selected has a strong effect on the estimated yield stress for a given tensile stress.

Finally, hardness measures were not available for all the results plotted in Figure 3.18 and reflected in Table 3.3. Figure 3.20 and Table 3.4, in contrast, represent only those tensile and yield stress samples for which hardness was also available. Like Figure 3.18, Figure 3.20 traces the estimated tensile-to-yield stress relationship. Like Table 3.3, Table 3.4 documents the lower tolerance bound yield stress for a range of tensile stresses. A more complete table of values is included in Appendix A. Overall, the two sets of results are comparable, confirming that the additional data do not bias the estimated relationship.

3.2 Hardness-to-Tensile Stress

As shown in Figures 3.9 and 3.13, neither plant versus transmission line piping nor decade of manufacture seemingly impacted the hardness-to-tensile relationship. In fact, none of the theoretical or empirical factors considered in Section 3.1 proved relevant. However, since distinct relationships were observed between ultimate tensile stress and yield stress based on plant piping versus transmission piping and based on decade of manufacture, these two portions of the data set were excluded from the final analysis.

The estimated hardness-to-tensile stress relationship is portrayed in Figure 3.21 by the solid line superimposed on the data. The parameters of the regression line are noted in the footnote to Figure 3.21. An asymptotic lower 99 percent confidence bound on the 0.5 percentile tensile stress is also given (dashed line) for each hardness. For a given hardness, therefore, at most 0.5 percent (with 99 percent confidence) of the transmission pipe lengths from heats manufactured prior to 1980 have tensile stresses below that represented by the dashed line. The data used to estimate this relationship and lower tolerance bound are summarized in Table 3.5.

In deriving the tolerance bound shown in Figure 3.21, the scatter about the line was assumed to follow a Normal (or Gaussian) distribution. Figure 3.22 presents a probability plot assessing the viability of this assumption. If the data are inconsistent with a Normal distribution, the plotted points will trend off the superimposed line. The limits of the probability plot represent how well the extremes of the data fit the assumed tails of the Normal distribution (e.g., a value of 2.57 represents the 0.5 percentile). A review of Figure 3.22 suggests the assumption is reasonable, though more so for the heart of the data's distribution than its limits. Given the problematic fit to the distribution's edge (tails), it is not surprising that a Shapiro-Wilks W statistical test rejects ($p < 0.0001$) the assumption.

Table 3.6 reports the lower tolerance bound tensile stress for a range of hardness. Since the confidence level and percentile to be used in any testing standard are still to be determined, Table 3.6 presents results for nine different combinations of confidence level (99.9 percent, 99 percent, and 95 percent) and percentile (0.1 percent, 0.5 percent, and 1 percent). A more thorough table of values is included in Appendix A. The specific combination selected has a profound effect on the estimated tensile stress for a given hardness.

3.3 Hardness-to-Yield Stress

As evidenced in Figures 3.10 and 3.14, neither plant versus transmission line piping nor decade of manufacture seemingly impacted the hardness-to-yield relationship. In fact, none of the studied theoretical or empirical factors described in Section 3.1 proved relevant. However, since distinct relationships were observed between ultimate tensile stress and yield stress based on plant piping versus transmission piping and based on date of manufacture, these two portions of the data set were excluded from the final analysis. Since this method was ultimately used for determining the yield stress, plots showing the effect of plant versus transmission pipe and decade of manufacture on the regression line are shown in Figures 3.23 and 3.24. In the case of date of manufacture, there are not enough data to estimate some of the regression lines accurately.

The estimated hardness-to-yield stress relationship is portrayed in Figure 3.25 by the solid line superimposed on the data. The estimated parameters of the line are noted in a footnote to Figure 3.25. An asymptotic lower 99 percent confidence bound on the 0.5 percentile yield stress is also traced (dashed line) for each hardness. For a given hardness, therefore, at most 0.5 percent (with 99 percent confidence) of the transmission pipe lengths from heats manufactured prior to 1980 have yield stresses below that represented by the dashed line. The data used to estimate this relationship and lower tolerance bound are summarized in Table 3.7.

In deriving the tolerance bound portrayed in Figure 3.25, the scatter about the line was assumed to follow a Normal (or Gaussian) distribution. Figure 3.26 presents a probability plot assessing the viability of this assumption. If the data are inconsistent with a Normal distribution, the plotted points will trend off the superimposed line. The limits of the probability plot represent how well the extremes of the data fit the assumed tails of the Normal distribution (e.g., a value of 2.57 represents the 0.5 percentile). A review of Figure 3.26 suggests the assumption is quite reasonable and the assumption is not rejected by a Shapiro-Wilks W test statistic.

Table 3.8 reports the lower tolerance bound yield stress for a range of hardness. Since several confidence levels and percentiles could be used in any testing standard, Table 3.8 presents results for nine different possible combinations of confidence level (99.9 percent, 99 percent, and 95 percent) and percentile (0.1 percent, 0.5 percent, and 1 percent). A more complete table of values is included in Appendix A. The specific combination selected has a strong effect on the estimated yield stress for a given hardness.

3.4   Yield-to-Tensile Stress Ratio versus Hardness

As noted in Section 1.1, current regulatory requirements require extensive use restrictions (including a maximum operating hoop stress of 6,000 psi) if the yield-to-tensile stress ratio (Y/T) of any of the sampled coupons equals or exceeds 0.85. Figure 3.27 plots the yield-to-tensile stress ratio versus hardness. Because of the tensile to yield stress relationship differences discussed in Section 3.1, only transmission line samples manufactured prior to 1980 are plotted in Figure 3.27. The estimated linear relationship between yield-to-tensile ratio and hardness is superimposed on the figure. Note that a significant portion of the data has a Y/T gradient that is greater than 0.85. Since much of these data comes from "in-service" pipe, the implication is that a Y/T limit of 0.85 is unreasonable. For this reason, the yield-to-tensile ratio reference line presented in Figure 3.27 is at 0.93, consistent with the current upper bound in API Specification 5L.

The objective in characterizing the yield-to-tensile-strength ratio relationship versus hardness is to estimate a maximum hardness associated with steel evidencing a yield-to-tensile ratio at or less than 0.93. This can be accomplished by estimating a tolerance bound on the relationship as was utilized in characterizing the relationships among yield stress, tensile stress, and hardness. In this instance, however, the objective is an upper tolerance bound estimate, with some degree of confidence, of the hardness value at which only a small percentile of yield-to-tensile stress ratios associated with the value exceed 0.93. If the hardness test were to estimate a hardness value in excess of this maximum value, then restrictions on the estimate of the maximum yield stress would be appropriate.

Since several confidence levels and percentiles could be used in any testing standard, nine different combinations of confidence level (99 percent, 95 percent, and 90 percent) and percentile (0.5 percent, 1 percent, and 5 percent) are considered in Table 3.9. The specific combination selected has a profound effect on the estimated maximum hardness value. Table 3.9 reports the hardness value at which the specified upper tolerance bound equals 0.93.

18

3.5   Recommended Data Analysis Process

Based on the developed relationships outlined above, yield stress can be estimated from hardness directly as well as indirectly through tensile stress. Figure 3.28 compares the direct method for determining yield stress and the indirect method. It is evident that there is a penalty in using the indirect method that is related to the statistical uncertainty of having two relationships versus a single direct relationship. Therefore, the direct approach is more viable and is adopted.

The confidence level and percentile to be used in any testing standard are still to be determined. Based on the available data, it is recommended here (and is reflected in the produced figures and tables) that a 99 percent lower confidence bound on the 0.5 percentile be the test statistic for the hardness testing. The selected percentile is based on what available data suggest is the current level of risk associated with newly manufactured pipe. Tensile testing from one manufacturer reported measured yield stresses below 52,000 psi for 2 out of 394 (or 0.51 percent) already qualified X52 pipe lengths. A histogram of these data is presented in Figure 3.29. Estimating a 99 percent lower confidence bound on this percentile further enhances the represented safety. Table 3.10 contains the hardness-to-yield stress conversions based on these recommendations. This table reflects an upper bound on the hardness values due to the upper bound of 0.93 of the Y/T ratio.

4.0 FIELD HARDNESS MEASUREMENT

As set out in Section 1.3, the approach that was adopted considered two relationships. One relationship involves yield and/or ultimate stress in terms of laboratory hardness. To this point in the report, only the laboratory-measured hardness relationship has been considered. The second involves the relationship between laboratory- and field-measured hardness. This section develops the relationship between laboratory and field measures of hardness.

Theoretically, any proven field-hardness measurement equipment could be used as long as the relationship between its measurements and Rockwell B laboratory hardness is or could be suitably established. Three commercially available hardness testers were evaluated as part of this project. Only the Equotip© Portable Metal Hardness Tester was further evaluated for the illustration later in this section. The other two testers had a high degree of variability in the "field" hardness results. It is possible that, with a higher degree of operator training or some refinement specific to measuring in situ hardness, this variability could be reduced.

Figure 4.1 shows the results developed using the three field measurement schemes. These hardness testers used a dynamic impact, ultrasonics, and a modified indentation test to determine hardness. No particular type of equipment is recommended by this report. However, as one field scheme must be adopted to illustrate the field data evaluation process, the Equotip© equipment was selected for this purpose. A generic procedure also is presented for correctly establishing the field-to-laboratory hardness relationship for any field hardness measurement technology under consideration. Given the necessarily generic character of this procedure, it may be beneficial to read the procedure and Equotip© illustration, and then re-read the generic procedure.

4.1 Procedure for Establishing a Field-to-Laboratory Hardness Relationship

The effort required to establish a relationship between field and laboratory hardness will likely vary considerably from one technology to another. The data and statistical analyses used

20

to establish a field-to-laboratory hardness measure relationship should mirror the data and statistical analyses described above for establishing the hardness-to-yield strength relationship. The procedure is presented here as a series of data collection or analytical steps representing a recipe for establishing a field-to-laboratory hardness relationship.

1. *Establish field-testing protocol.* — Before data collection can be initiated, a comprehensive field testing protocol must be developed/identified for the field hardness technology under consideration. What surface preparation measures are required by the technology? What protocol steps can be included to reduce operator-to-operator differences? Is operator training required, for example? How frequently should the testing apparatus be calibrated? The objective in answering such questions is to formulate a protocol that reduces, to the extent possible, any extraneous variability associated with the field hardness measurement process. In addition, the protocol should acknowledge any testing conditions (is there, for example, an ambient temperature range within which samples must be collected?) required by the candidate technology.

2. *Collect paired field and laboratory hardness samples.* — The procedure begins with the collection of data suitable for assessing the relationship between the field hardness measurement technology under consideration and laboratory-measured Rockwell B hardness. At minimum, 25 to 50 paired samples using both measures (e.g., the hardness of a pipe length is measured in the laboratory and using the candidate technology) on pipe lengths from a broad range of relevant manufacturers and heats are necessary. Visualize these data as two columns of hardness measurements with each row representing a sampled pipe length. Effectively, the collected samples should reflect the pipe for which the relationship, once established, will be applied. The greater the number of samples, the tighter the statistical bound estimated for the relationship is to the targeted percentile laboratory hardness.

The collected samples also should reflect variability in whatever factors might impact the measurements. If, for example, wall thickness impacts the candidate measurement technology, the collected data should include paired samples collected from pipe lengths of relevant wall thickness. If operating pressure influences the field hardness measurements, its effect must be reflected in the measurement variability. Factors that are relevant for in-service pipe lengths are particularly important to study, as it may be possible to establish distinct relationships at various factor values (see item #4 below). Additional data columns (one for each factor) recording the relevant factor value are necessary if these relationships are to be assessed later.

Finally, the observed field hardness values should cover the span of values that might be expected. Extrapolating the relationship beyond the data is technically feasible, but requires the questionable assumption that the established relationship is maintained in this domain.

3. *Analyze relationship(s) between field and laboratory hardness.* — Once suitable data are available, the associated relationship between field and laboratory measured hardness must be characterized. Visually the relationship is evident for the collected data as a field versus laboratory hardness scatterplot. Such a relationship is not necessarily linear; non-linear relationships are entirely plausible. Linear or non-linear regression on the collected data should be used to estimate the parameters of the assumed relationship.[1] Before non-linear regression can be employed, of course, the form of the non-linear relationship must be established. For example, if laboratory hardness plateaus at higher values of field hardness, the assumed equation embodying the relationship should show such a plateau. In most circumstances, however, a linear relationship is probably a reasonable assumption.

A number of statistical computations exist to assess whether a linear relationship is appropriate — e.g., a test of lack of fit versus pure error[2], a test of the statistical significance of the quadratic term in a fitted quadratic relationship. The relative merits of these computations depend somewhat on the specific circumstances being considered and

---

[1] Draper, N. R. and Smith, H., **Applied Regression Analysis**, Wiley Series in Probability and Mathematical Statistics, John Wiley and Sons, Inc.
[2] Ibid, page 33.

the statistical hypothesis being tested. Inspection of the scatterplot, however, is often sufficient to determine the appropriateness (or lack thereof) of a linear relationship.

4. *Characterize effect of identifiable factors.* — Before statistically bounding the characterized relationship, the effect of each 'identifiable' factor influencing the relationship should be estimated. 'Identifiable' factors are those that can be readily measured or determined for in-service pipe lengths (e.g., wall thickness). The question to answer is the effect such factors have on the relationship. For example, does wall thickness impact the slope or intercept value of the linear relationship or the linear character to the relationship? In essence, the relationship for each subset of the 'identifiable' factor in question is characterized (see item #3). Doing so has advantages and disadvantages. For some of the wall thickness values, for example, the relevant relationship may indicate higher laboratory hardness for particular field hardness than was indicated by the common relationship estimated irrespective of wall thickness. Moreover, if the factor does indeed influence the field to laboratory hardness relationship, reduced scatter will be evident about the factor-specific characterized relationships. The advantage of this reduction will be manifest when a statistical bound is estimated (see item #5). The drawback to characterizing factor-specific relationships is that fewer data are associated with the characterization of a given relationship. The result is a widening of the statistical bound estimated for the relationship. The magnitude of this consequence depends on just how much data were collected; establishing a statistical bound based on 250 rather than 500 paired samples, for example, has trivial consequence as compared to establishing it with 25 rather than 50 paired samples. As an alternative, it may be possible to determine an equation that fits the data and then bound that equation.

5. *Estimate lower tolerance bound on laboratory hardness.* — Once the collection (recognizing the effect of 'identifiable' factors) of field-to-laboratory hardness relationships has been characterized, a lower tolerance bound on each relationship must be estimated. Specifically, a lower 99 percent confidence bound on the 0.5 percentile laboratory hardness must be estimated for each relevant relationship. Visually this amounts to bounding the exhibited scatter about the relationship. Such a statistical bound

can be derived for both linear and non-linear relationships. Suitably derived bounds for non-linear relationships, however, are particularly conservative as the degree of non-linearity is necessarily among the factors contributing to the uncertainty reflected in the estimated bound.

In estimating a lower tolerance bound on yield strength as a function of laboratory hardness, a Normal (Gaussian) distribution is assumed for the scatter about the linear relationship. The assumption of normally distributed scatter about a characterized relationship is often reasonable (at least asymptotically so), but should be evaluated by generating and reviewing probability plots like that presented in Figure 3.19.[3] If the scatter has a skewed character about the estimated relationship, the Lognormal distribution is sometimes appropriate. This skewed distribution is readily applied by reviewing the log-transformed laboratory hardness measures versus field hardness. Other skewed distributions such as Weibull or Extreme Value may be more appropriate though the associated computations to estimate a lower tolerance bound are typically more complex and are supported by fewer statistical analysis software packages.[4] If a linear relationship is evident and a Normal distribution to the scatter can be assumed, a 99 percent lower confidence bound on the 0.5 percentile (or any other tolerance bound) can be readily calculated.[5] Once estimated, the lower tolerance bound on the field to laboratory hardness relationship is used to translate the minimum measured field hardness to laboratory hardness.

As when estimating yield strength, the specific confidence bound and percentile combination selected has a profound effect on the resulting estimate. The values for the two bounds should be set together since both will be used to translate the results of field hardness testing into estimated yield strength. Since the objective is to maintain, across both translations, an equivalent level of safety to current pipeline safety standards, a 99

---

[3] Ibid, page 143.
[4] Nelson, W., **Applied Life Data Analysis**, Wiley Series in Probability and Mathematical Statistics, John Wiley and Sons, Inc.
[5] Graybill, F. A., **Theory and Application of the Linear Model**, Wadsworth Publishing Company, Inc.

percent lower confidence bound on the 0.5 percentile laboratory hardness is recommended.

4.2 Illustration of Procedure

To illustrate the procedure for characterizing the relationship between laboratory hardness and a field hardness measurement technology, the Equotip© Portable Metal Hardness Tester[6] manufactured by Corvib Canada was utilized. Its use does not represent a recommendation or an implicit criticism of other sampling technologies.

1. *Establish field-testing protocol.* — Evaluation of the Equotip© Tester indicated operator training beyond general familiarity with the apparatus was not necessary. Three operators examined each of 40 pipe lengths, collecting ten samples at each of three circumferential locations on each length. On one of the pipe lengths, four axial locations on the length were sampled each of 10 times. An analysis of variance (ANOVA) of the resulting variability in measured hardness suggested operator differences contribute minimally to the overall measurement variability. Figure 4.2 presents stacked histogram bars reporting the percentile of variability attributable to operator, location (axial or circumferential), and random unknown factors.

    The field-testing protocol for Equotip© testing, therefore, consisted primarily of surface preparation procedures. In particular, preparation of the pipe length's surface began by rough grinding the surface first using 60 and then 80 grit material. A sander with a 240-grit disk was then used, followed by a sander with a 400-grit disk for final polish. Scratch marks from the previous step were removed in each new step. These procedures were developed as part of an earlier application of the Equotip© Tester.

2. *Collect paired field and laboratory hardness samples.* — Data collection using the Equotip© Tester entailed ten measurements collected at a single site (after surface preparation) from each of 40 pipe lengths for which laboratory hardness measures were

---

[6] <http://www.corvib.com/equotip/equotip1.htm>

also available. The 40 pipe lengths studied represented a broad range of steel grades, pipe diameters, wall thickness, and manufacturers. All the samples were collected on unpressurized rings of pipe. Work by Columbia Gas Transmission indicated operating pressure has little effect on Equotip© Tester-derived field measures of hardness, while wall thickness had a significant effect.

3. *Analyze relationship(s) between field and laboratory hardness.* — The relationship between field and laboratory hardness suggested by the collected data and estimated using linear regression is presented in Figure 4.3 (solid line). The plotted data include that used to assess operator- and location-induced variability (see item #1). A linear relationship seems a reasonable assumption, though statistical tests suggest mixed evidence. A review of the data scatter evident in Figure 4.3 does not suggest an alternative form to the relationship, so a linear model was used in this illustration.

4. *Characterize effect of identifiable factors.* — It is known that wall thickness impacts Equotip© Tester-derived field measures of hardness. However, there are insufficient data at this time for statistical analysis. Since wall thickness does have an effect, distinct field-to-laboratory hardness relationships should be developed for each distinct wall thickness. There are insufficient data available in this illustration to develop these relationships, but the approach is straightforward. An alternative approach would be to determine a mathematical relationship that incorporates the wall thickness and find a lower bound for this relationship. Plots like Figure 4.3 are examined, though the plotted scatter and fitted relationship consider only those data representing pipe lengths of a given wall thickness. These plots should suggest relationships whose parameters can be estimated using linear or non-linear regression and evaluated using relevant statistical analyses.

5. *Estimate lower tolerance bound on laboratory hardness.* — A lower tolerance bound on the field-to-laboratory hardness relationship is shown in Figure 4.4. Specifically, the solid line in Figure 4.4 traces a lower 99 percent confidence bound on the 0.5 percentile laboratory hardness for a given field hardness. Had sufficient data been available to develop the effect of wall thickness, distinct bounds would be portrayed for each relevant

wall thickness. Table 4.1 reports, for a range of field hardness, the lower tolerance bound laboratory hardness derived using nine different combinations of confidence level (99.9 percent, 99 percent, and 95 percent) and percentile (0.1 percent, 0.5 percent, and 1 percent).

For example, say that the field hardness measurements resulted in a hardness of 90. If a lower 99 percent confidence level on the 0.5 percentile is used, then the laboratory hardness is found to be 81 from Table 4.6. Referring to Table A.3.8 on page A-22, this results in a yield strength of 42,100 psi.

5.0 SAMPLING METHODOLOGY

With relationships developed between field hardness, laboratory hardness, and yield and ultimate tensile stress, a sampling methodology is needed to implement them. This procedure is for long pipelines where many samples would be required. For shorter pipe runs, it will be necessary to set minimum sample requirements such as those in 49 CFR Part 192, Appendix B.

The sampling frequency and design in the current Federal requirements (i.e., one out of every ten lengths for lines with more than ten lengths) implies a specific degree of uncertainty. Sequential lengths within a pipeline are not necessarily from the same heat; in fact, the lengths from each heat used in the pipeline's construction could be almost randomly distributed throughout the pipeline, depending on shipping practices. By sampling one out of every ten sequential lengths, there is an associated possibility of not sampling at least one length from each heat represented in the pipeline. By choosing to collect from each sampled length a single characterization of tensile properties (i.e., a single coupon mill testing), heat-to-heat variability is presumed to exceed any axial or circumferential variability with a given pipe length. This assumption, it should be emphasized, is consistent with the data reported here. Since the sampling will include some heats repeatedly, within-heat variability is captured to some extent.

To assess the uncertainty of not sampling from all a pipeline's heats, analyses were conducted to estimate the associated probability of missing heats. Statistical simulation was used to mimic the current sampling methodology. After each iteration, the number of heats was calculated for which none of its constituent lengths were sampled. Because each heat's lengths were assumed to be randomly distributed throughout the pipeline for this analysis, the relative frequency (or probability) of sampled (as opposed to "missed") heats is independent of pipeline length; the longer the pipeline, the larger the number of represented heats. To calculate the relative frequency of sampled heats, the simulations assumed 50-ton heats of standard wall thickness for a range of pipe diameter. This heat size is representative of the heat sizes used for manufacturing the pipe during the period relevant to this study. Current heat sizes can be much larger. Table 5.1 reports, for each assumed pipe diameter, the estimated number of lengths for each 50-ton heat.

Figure 5.1 presents the estimated relative frequency of sampled heats versus sampling frequency assuming pipe diameters of 12, 16, 20, 24, 30, 36, and 42 inches with standard wall thickness from API Specification 5L. Smaller sampling frequencies prompt decreases in the relative frequency of sampled heats — exponential in character over the 1 in every 10 to 100 domain — and larger pipe diameters produce lower relative frequencies of heats.

Whereas the current requirements assume a constant sampling frequency across all pipe diameters, it is recommended instead that the sampling frequency be statistically adjusted as a function of pipe diameter while retaining the same relative frequency of sampled heats evident for 42-inch diameter pipe. Table 5.2 reports the recommended sampling frequency at each pipe diameter. A review of Figure 5.1 indicates the origin of the values cited in Table 5.2. In Figure 5.1, the current sampling frequency of one in ten for 42-inch diameter pipe has a relative frequency of sampled heats equal to 0.986, while for 16-inch diameter pipe the same relative frequency of sampled heats is achieved with a sampling frequency of one in 67. Wall thickness also impacts the relative frequency of sampled heats. More heats will be required of pipelines with thicker walls but the same pipe diameter, which in turn will increase the number of samples.

It is important to note that the uncertainty associated with sampled heats is actually considerably higher than the relative frequency estimated here. How much higher depends upon the actual distribution of tensile properties across manufactured heats. The probability of missing sampled heats derives from the potential for missing heats with unusually low tensile properties (i.e., in the lower tail of the distribution). However, the heats missed could show tensile properties similar to if not higher than those sampled. The uncertainty, therefore, depends upon the probability of a heat with unusually low tensile properties as well as the relative frequency of sampled heats.

Available data support the implicit assumption in current requirements that heat-to-heat differences represent the principal source of variability in a pipeline's tensile properties. As such, collection of multiple hardness samples at different axial or circumferential locations on a sampled length is not recommended. These additional samples cannot take the place of samples from other pipe lengths.

6.0 CONCLUSIONS/RECOMMENDATIONS

1. Statistically based procedures have been developed to determine the yield stress of transmission piping of grades X52 or less. The determination of yield stress can be accomplished either directly from hardness measures or indirectly by relating these measures to tensile stress, which in turn is related to yield stress. The first or direct procedure is recommended because it minimizes uncertainty and so translates into higher yield stress determinations for a given hardness (i.e., statistical uncertainty needs be accommodated for only one relationship rather than two). Statistical uncertainty in the determination is accounted for by estimating a lower tolerance bound of the relationship between yield stress and hardness. Specifically, it is recommended that a 99 percent lower confidence bound on the 0.5 percentile yield stress be estimated for a given hardness. **Since statistical methods are used to determine a lower bound, the resulting yield strength will be lower than the SMYS yield strength.**

    Included in the developed procedures is an upper bound laboratory hardness value of 87 HRB that is consistent with a derived bound on the yield-to-tensile ratio of the pipe equaling or exceeding 0.93. Specifically, 87 HRB is the hardness value at which a 99 percent upper confidence bound on the 99 percentile yield-to-tensile-strength ratio equals 0.93. The decision to use 0.93 as the maximum yield-to-tensile ratio is a reflection of the currently acceptable upper bound on yield-to-tensile ratio specified in API 5L.

2. Any hardness measurements of buried pipe will most likely not produce the same established laboratory measurements of hardness that were used to develop the relationship between hardness and yield stress. It is necessary, therefore, to estimate laboratory hardness using the collected field hardness data. The relationship necessary to accomplish such estimation will need to be developed — including incorporation of statistical uncertainty in the relationship — for any field test procedure being considered. The procedures documented in this report recommend

that the relationship be characterized by a 99 percent lower confidence bound on the 0.5 percentile laboratory hardness for a given field hardness.

3. The current sampling procedure in CFR 192 Appendix B effectively requires tensile properties be measured for one of every ten pipe lengths. The developed statistical procedures instead utilize pipe diameter to determine how frequently tensile properties should be characterized. Statistical simulation indicated such an approach produces the same degree of uncertainty in sampling all heats represented in the pipeline under consideration.

4. The statistical procedures developed in this project maintain the degree of uncertainty evidenced by current testing requirements. The limited available data indicate the lower 0.5 percentile of a manufacturer's heat labeled X52 actually have measured yield stress below 52,000 psi. Also, the probability of failing to sample all heats is consistent with the relative frequency of sampled heats estimated for 42-inch diameter pipe.

Table 2.1. Sample frequency and percentile by decade of manufacture among collected data with tensile property measures

Decade	# of Samples	Percent of Samples
30's	13	0.3
40's	15	0.3
50's	153	3.3
60's	628	13.4
70's	327	7.0
80's	60	1.3
90's	67	1.4
Unknown	3,437	73.1

Table 2.2. Sample frequency and percentile by steel grade among collected data with tensile property measures

Grade	# of Samples	Percent of Samples
290CI	9	0.2
290CII	8	0.2
359CI	23	0.5
359CII	24	0.5
46CII	36	0.8
46CIII	2	0.0
52CII	178	3.8
52CIII	1	0.0
A	17	0.4
A106A	4	0.1
A106B	27	0.6
A25	2	0.0
A30	3	0.1
A53	37	0.8
A53B	1,970	41.9
B	100	2.1
Unknown	942	20.0
X40	2	0.0
X42	138	2.9
X45	3	0.1
X46	112	2.4
X48	1	0.0
X50	10	0.2
X52	1,051	22.4

Table 2.3. Sample frequency and percentile by hardness measure among collected data with tensile and hardness property measures

Hardness Measure	# of Samples	Percent of Samples
BHN	22	2.6 %
DPH	26	3.1 %
HRB	771	92.2 %
HRC	1	0.1 %
KHN	16	1.9 %

Table 3.1. Fitted linear models to date restricted by pipe type and by decade of manufacture

Collected Data Considered	Parameters of Fitted Linear Regression		
	Intercept	Slope	Root Mean Square Error
Overall	15,085	0.582	4,798
Plant Piping	-8,851	0.939	3,852
Transmission	15,334	0.556	4,426
30's	12,151	0.488	3,763
40's	11,948	0.576	6,010
50's	-4,070	0.779	3,155
60's	2,930	0.705	4,679
70's	26,125	0.445	4,537
80's	520	0.783	4,743
90's	11,123	0.704	4,987
Unknown Decade	13,523	0.608	4,587

Table 3.2. Summary statistics on yield strength versus ultimate tensile strength among collected data with tensile property measures

Ultimate Tensile Strength							
Mean	Standard Deviation	Percentile					
		5th	10th	25th	75th	90th	95th
77,238	7,854	65,200	67,500	71,600	82,740	89,710	95,300

Yield Strength							
Mean	Standard Deviation	Percentile					
		5th	10th	25th	75th	90th	95th
58,102	6,057	47,270	50,900	55,300	61,760	64,800	66,900

Pearson's Correlation Coefficient	Root Mean Square Error of Linear Fit
0.52	4,217

Table 3.3. Estimated lower tolerance bound yield strength as function of measured ultimate tensile strength, confidence level, and percentile - based on collected data with tensile property measures for transmission line lengths manufactured prior to 1980

Targeted Percentile	Confidence Level	Measured Ultimate Tensile Strength (PSI)	Estimated Lower Tolerance Bound on Yield Strength (PSI)
0.1%	99.9%	50,000	28,500
		60,000	34,200
		70,000	39,800
		80,000	45,400
		90,000	50,800
		100,000	56,200
	99.0%	50,000	28,900
		60,000	34,500
		70,000	40,100
		80,000	45,700
		90,000	50,600
		100,000	56,600
	95.0%	50,000	29,200
		60,000	34,800
		70,000	40,400
		80,000	45,900
		90,000	51,500
		100,000	56,900
0.5%	99.9%	50,000	30,800
		60,000	36,500
		70,000	42,200
		80,000	47,700
		90,000	53,200
		100,000	58,500
	99.0%	50,000	31,100
		60,000	36,800
		70,000	42,500
		80,000	48,000
		90,000	53,500
		100,000	58,800
	95.0%	50,000	31,400
		60,000	37,100
		70,000	42,700
		80,000	48,200
		90,000	53,700
		100,000	59,100
1.0%	99.9%	50,000	31,900
		60,000	37,600
		70,000	43,300
		80,000	48,900
		90,000	54,300
		100,000	59,600
	99.0%	50,000	32,200
		60,000	37,900
		70,000	43,500
		80,000	49,100
		90,000	54,600
		100,000	60,000
	95.0%	50,000	32,500
		60,000	38,200
		70,000	43,800
		80,000	49,300
		90,000	54,800
		100,000	60,200

Table 3.4. Estimated lower tolerance bound yield strength as function of measured ultimate tensile strength, confidence level, and percentile — based on collected data with tensile and hardness property measures for transmission line lengths manufactured prior to 1980

Targeted Percentile	Confidence Level	Measured Ultimate Tensile Strength (PSI)	Estimated Lower Tolerance Bound on Yield Strength (PSI)
0.1%	99.9%	50,000	23,200
		60,000	30,400
		70,000	37,400
		80,000	44,100
		90,000	50,500
		100,000	56,800
	99.0%	50,000	23,800
		60,000	30,900
		70,000	37,800
		80,000	44,500
		90,000	51,000
		100,000	57,400
	95.0%	50,000	24,300
		60,000	31,300
		70,000	38,200
		80,000	44,900
		90,000	51,500
		100,000	57,900
0.5%	99.9%	50,000	25,900
		60,000	33,000
		70,000	40,000
		80,000	46,800
		90,000	53,200
		100,000	59,400
	99.0%	50,000	26,300
		60,000	33,400
		70,000	40,400
		80,000	47,100
		90,000	53,600
		100,000	60,000
	95.0%	50,000	26,800
		60,000	33,800
		70,000	40,700
		80,000	47,500
		90,000	54,000
		100,000	60,500
1.0%	99.9%	50,000	27,000
		60,000	34,300
		70,000	41,300
		80,000	48,100
		90,000	54,400
		100,000	61,200
	99.0%	50,000	27,600
		60,000	34,700
		70,000	41,700
		80,000	48,400
		90,000	54,900
		100,000	61,200
	95.0%	50,000	28,100
		60,000	35,000
		70,000	42,000
		80,000	48,800
		90,000	55,300
		100,000	61,700

Table 3.5. Summary statistics for ultimate tensile strength versus Rockwell B hardness among collected data with tensile and hardness property measures

Hardness							
Mean	Standard Deviation	Percentile					
		5th	10th	25th	75th	90th	95th
85.6	5.12	77	81	84	88.1	90.5	91.7

Ultimate Tensile Strength							
Mean	Standard Deviation	Percentile					
		5th	10th	25th	75th	90th	95th
75,616	7,740	65,000	67,200	69,960	81,425	86,000	87,500

Pearson's Correlation Coefficient	Root Mean Square Error of Linear Fit
0.23	6,778

Table 3.6. Estimated lower tolerance bound ultimate tensile strength as a function of measured Rockwell B hardness, confidence level, and percentile - based on collected data with tensile and hardness property measures for transmission line lengths manufactured prior to 1980

Targeted Percentile	Confidence Level	Measured Hardness (HRB)	Estimated Lower Tolerance Bound on Tensile Strength (PSI)
0.1%	99.9%	55	26,300
		65	35,200
		75	43,900
		85	51,900
		95	58,600
	99.0%	55	27,800
		65	36,300
		75	44,600
		85	52,500
		95	59,400
	95.0%	55	29,100
		65	37,300
		75	45,300
		85	53,000
		95	60,000
0.5%	99.9%	55	29,900
		65	38,800
		75	47,600
		85	55,700
		95	62,400
	99.0%	55	31,400
		65	39,900
		75	48,300
		85	56,200
		95	63,100
	95.0%	55	32,700
		65	40,900
		75	48,900
		85	56,700
		95	63,600
1.0%	99.9%	55	31,600
		65	40,600
		75	49,400
		85	57,500
		95	64,200
	99.0%	55	33,100
		65	41,700
		75	50,100
		85	58,000
		95	64,800
	95.0%	55	34,400
		65	42,600
		75	50,700
		85	58,400
		95	65,400

Table 3.7. Summary statistics for yield strength versus Rockwell B hardness among collected data with tensile and hardness property measures

Hardness							
Mean	Standard Deviation	Percentile					
		5th	10th	25th	75th	90th	95th
85.6	5.12	77	81	84	88.1	90.5	91.7

Yield Strength							
Mean	Standard Deviation	Percentile					
		5th	10th	25th	75th	90th	95th
58,092	5,745	49,800	53,340	55,600	61,000	64,930	67,000

Pearson's Correlation Coefficient	Root Mean Square Error of Linear Fit
0.38	4,512

Table 3.8. Estimated lower tolerance bound yield strength as a function of Rockwell B hardness, confidence level, and percentile - based on collected data with tensile and hardness property measures for transmission line lengths manufactured prior to 1980

Targeted Percentile	Confidence Level	Measured Hardness (HRB)	Estimated Lower Tolerance Bound on Yield Strength (PSI)
0.1%	99.9%	55	18,700
		65	26,700
		75	34,600
		85	42,100
		95	48,700
	99.0%	55	19,700
		65	27,500
		75	35,100
		85	42,500
		95	49,200
	95.0%	55	20,600
		65	28,100
		75	35,600
		85	42,800
		95	49,600
0.5%	99.9%	55	21,100
		65	29,200
		75	37,100
		85	44,600
		95	51,200
	99.0%	55	22,100
		65	29,900
		75	37,600
		85	45,000
		95	51,600
	95.0%	55	23,000
		65	30,500
		75	38,000
		85	45,300
		95	52,000
1.0%	99.9%	55	22,300
		65	30,400
		75	38,300
		85	45,800
		95	52,400
	99.0%	55	23,300
		65	31,100
		75	38,800
		85	46,200
		95	52,800
	95.0%	55	24,100
		65	31,700
		75	39,200
		85	46,400
		95	53,200

Table 3.9. Maximum Rockwell B hardness as a function of confidence level and percentile - based on collected data with tensile and hardness property measures for transmission line lengths manufactured prior to 1980

Targeted Percentile	Confidence Level	Hardness (HRB) at which upper bound on Y/T ratio exceeds 0.93
0.1%	99.9%	< 54
	99.0%	59
	95.0%	65
0.5%	99.9%	77
	99.0%	80
	95.0%	82
1.0%	99.9%	86
	99.0%	87
	95.0%	89

Table 3.10 Estimated lower tolerance bound yield strength as a function of Rockwell B hardness - based on collected data with tensile and hardness property measures for transmission line lengths manufactured prior to 1980

Targeted Percentile	Confidence Level	Measured Hardness (HRB)	Estimated Lower Tolerance Bound on Yield Strength (PSI)
0.50%	99.0%	55.4	24,000
		65.0	29,900
		75.0	37,600
		85.0	45,000
		87.0	46,300

Table 4.1. Estimated lower tolerance bound laboratory Rockwell B hardness as a function of measured field hardness, confidence level, and percentile - based on collected data with tensile and hardness property measures for transmission line lengths manufactured prior to 1980

Targeted Percentile	Confidence Level	Measured Field Hardness (HRB)	Lower Tolerance Bound of Lab Hardness (HRB)
0.1%	99.9%	55	55
		60	58
		65	62
		70	65
		75	68
		80	72
		85	75
		90	78
		95	81
	99.0%	55	55
		60	59
		65	62
		70	65
		75	69
		80	72
		85	75
		90	78
		95	82
	95.0%	55	55
		60	59
		65	62
		70	66
		75	69
		80	72
		85	75
		90	79
		95	82
0.5%	99.9%	55	57
		60	61
		65	64
		70	68
		75	71
		80	74
		85	77
		90	80
		95	84
	99.0%	55	58
		60	61
		65	64
		70	68
		75	71
		80	74

Targeted Percentile	Confidence Level	Measured Field Hardness (HRB)	Lower Tolerance Bound of Lab Hardness (HRB)
		85	78
		90	81
		95	84
	95.0%	55	58
		60	61
		65	65
		70	68
		75	71
		80	75
		85	78
		90	81
		95	84
1.0%	99.9%	55	58
		60	62
		65	66
		70	69
		75	72
		80	75
		85	79
		90	82
		95	85
	99.0%	55	59
		60	62
		65	66
		70	69
		75	72
		80	76
		85	79
		90	82
		95	85
	95.0%	55	59
		60	62
		65	66
		70	69
		75	72
		80	76
		85	79
		90	82
		95	85

Table 5.1. Number of lengths of given pipe diameter produced from 50 ton heat (assuming 0.332 inch wall thickness)

Pipe Diameter (inches)	# of Joints from 50 ton Heat
12	50
16	40
20	32
24	26
30	21
36	17
42	15

Table 5.2. Recommended sampling frequency, as function of pipe diameter, developed from simulation analyses examining the relative frequency of sampled heats

Pipe Diameter (inches)	Sampling Frequency (1 out of every ... joints)
12 or less	100
16	67
20	38
24	25
30	17
36	12
42	10

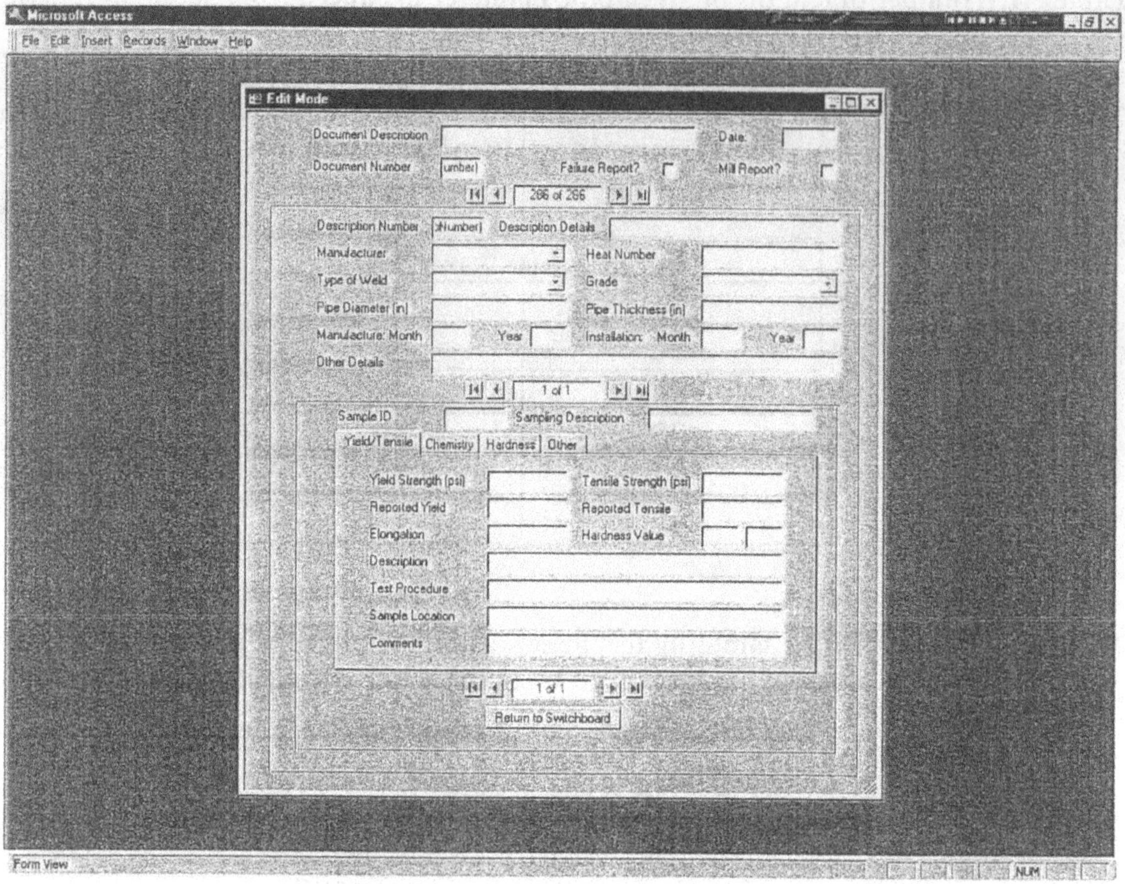

Figure 2.1. Main data entry/review form of project database used to store collected data

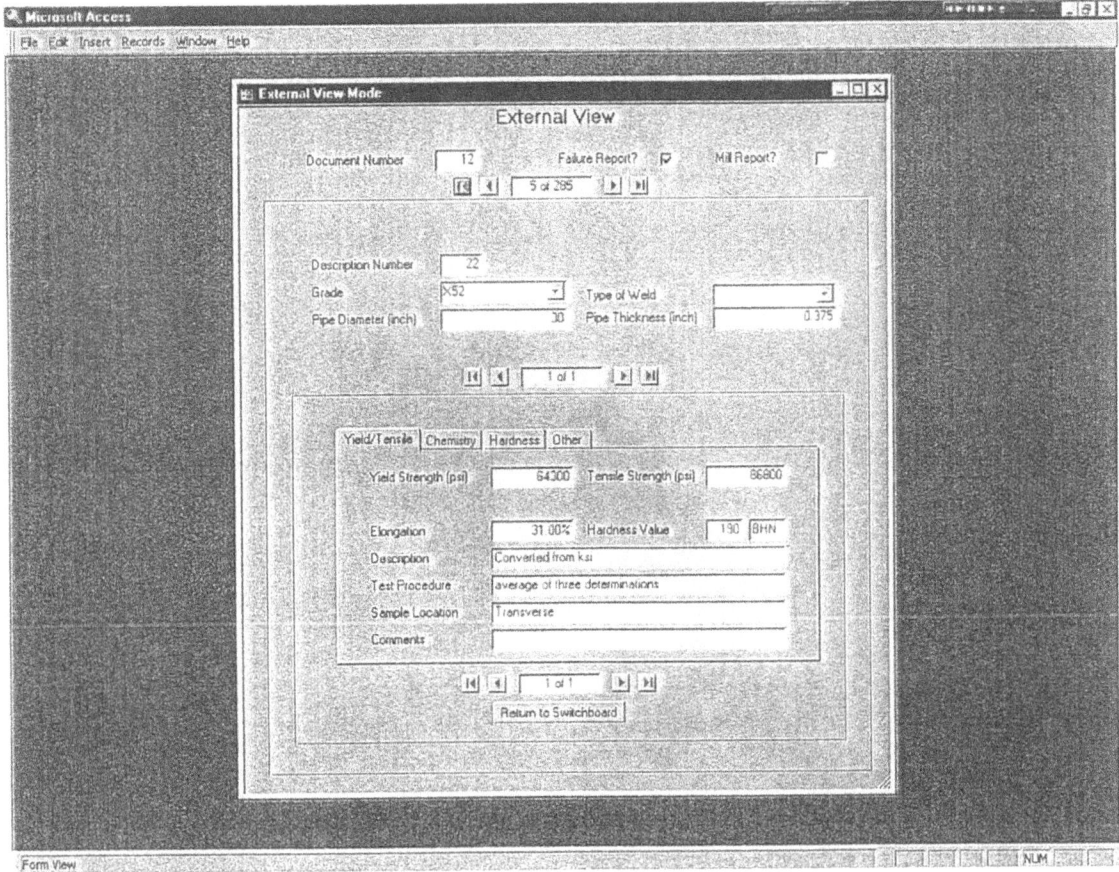

Figure 2.2. Main data entry/review form of sanitized project database used to store collected data

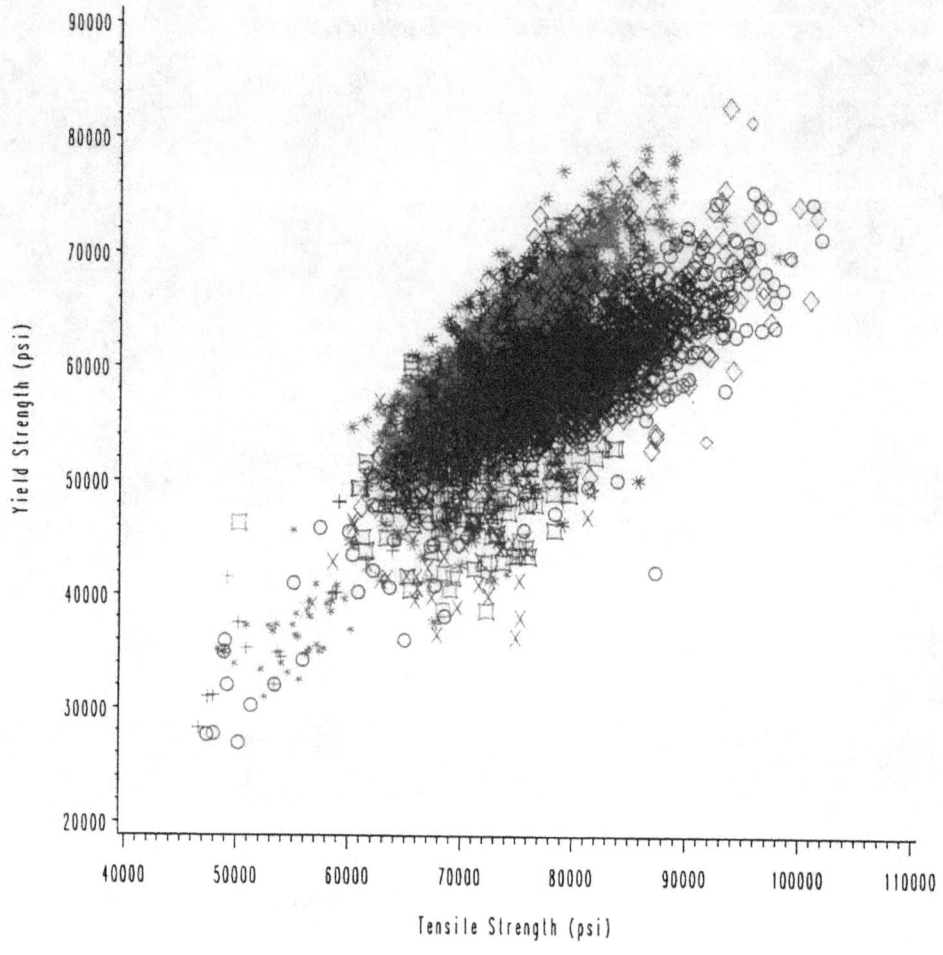

Figure 3.1. Yield strength versus ultimate tensile strength by steel grade (distinguished by unique symbol types) — collected data with tensile property measures

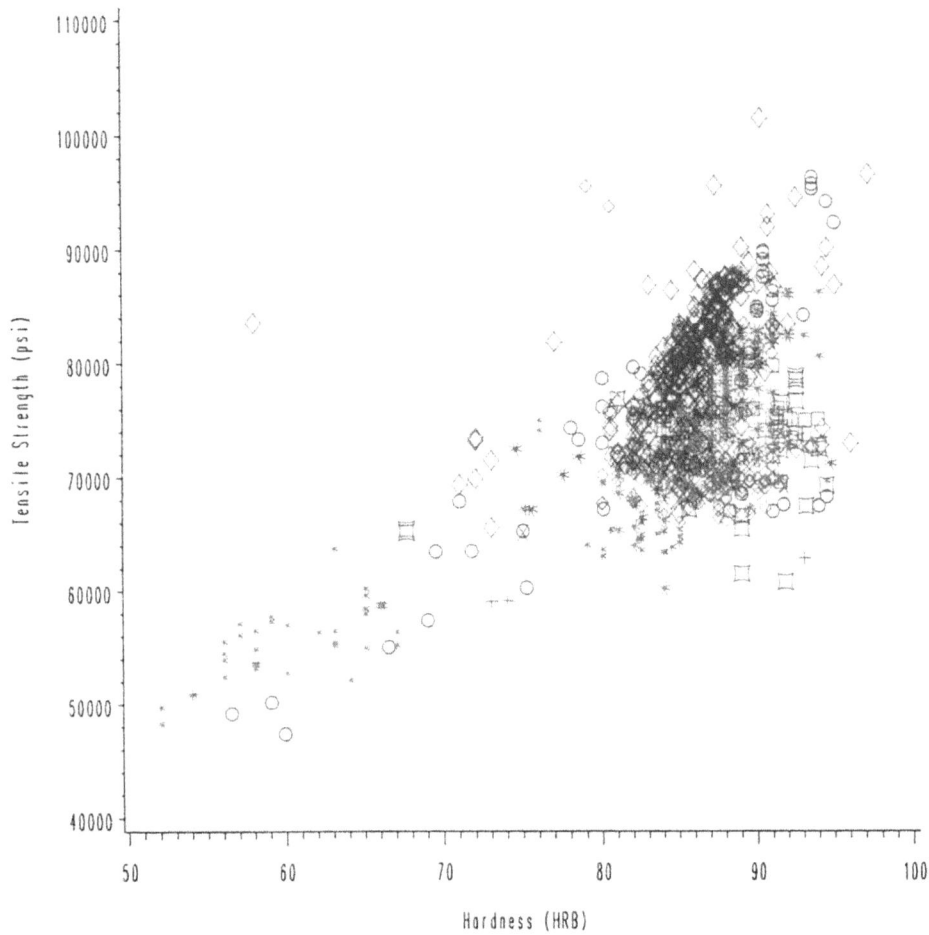

Figure 3.2. Ultimate tensile strength versus Rockwell B hardness by steel grade (distinguished by unique symbol types) — collected data with tensile and hardness property measures

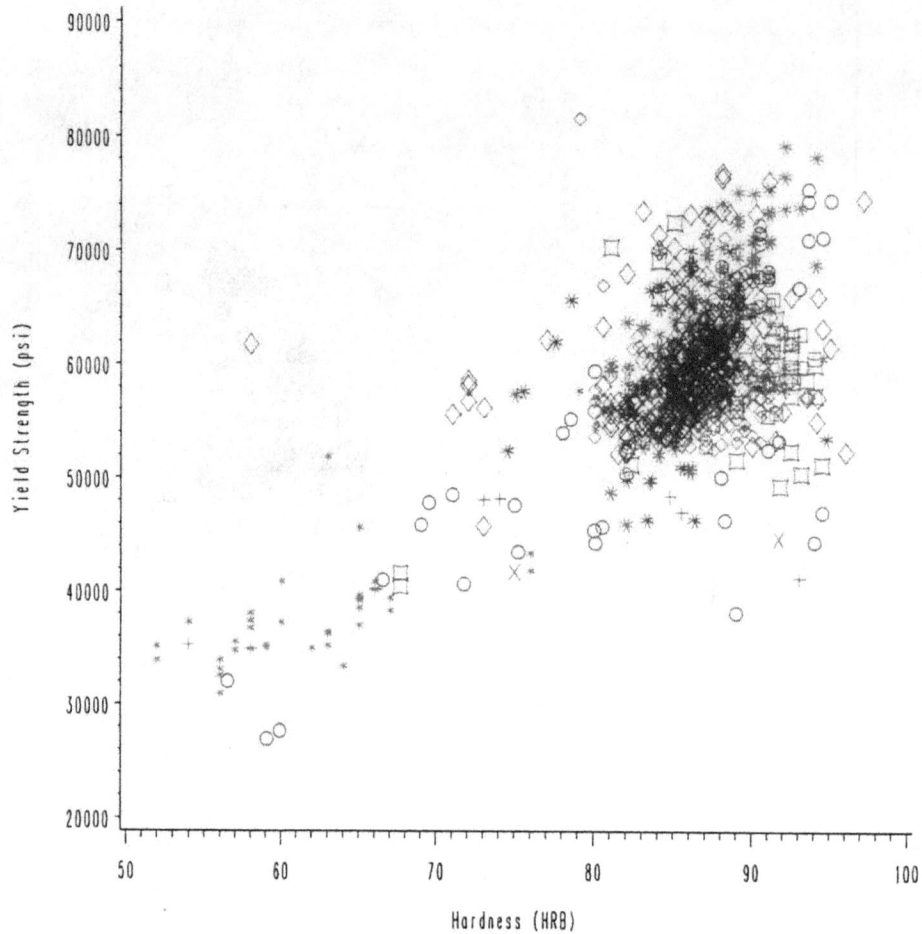

Figure 3.3. Yield strength versus Rockwell B hardness by steel grade (distinguished by unique symbol types) — collected data with tensile and hardness property measures

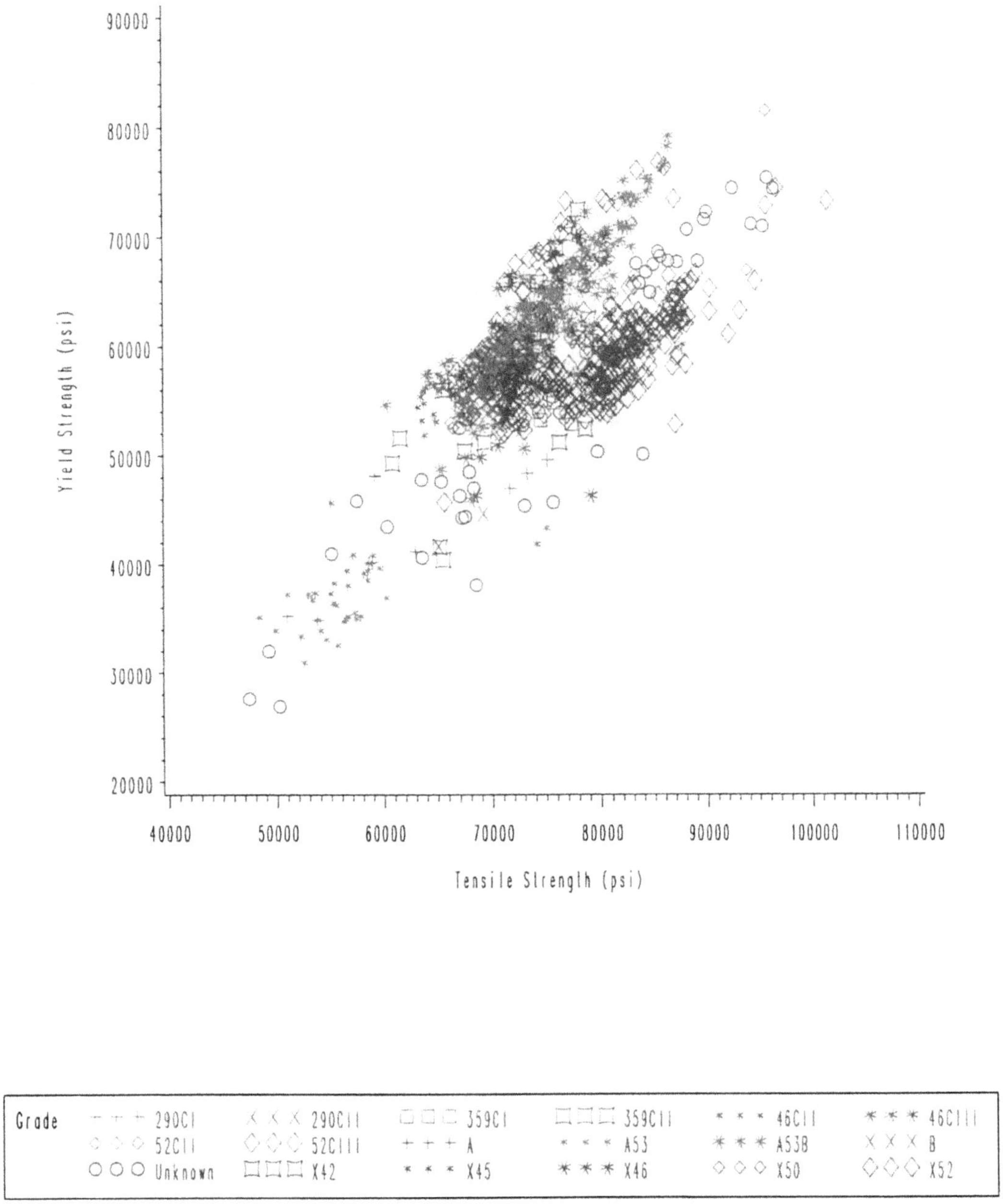

Figure 3.4 Yield strength versus ultimate tensile strength by steel grade (distinguished by unique symbol types) — collected data with tensile and hardness property measures

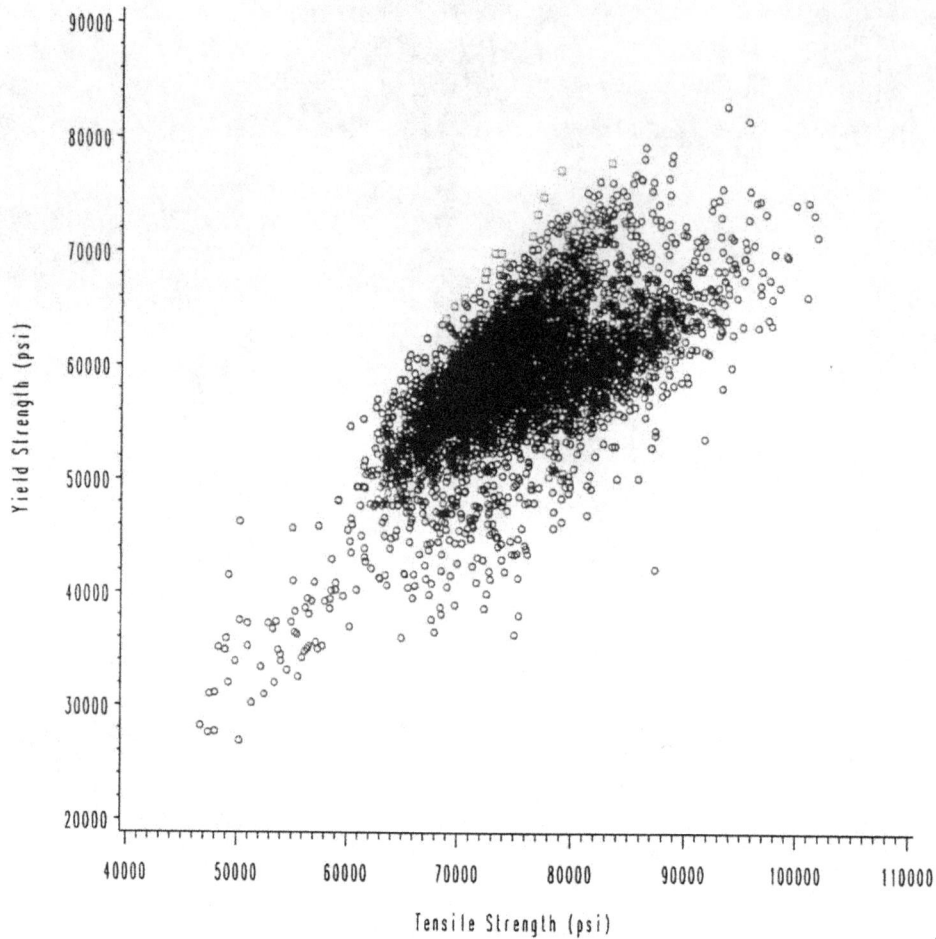

Figure 3.5. Yield strength versus ultimate tensile strength by yield-to-tensile ratio category (distinguished by unique symbol types) — collected data with tensile property measures

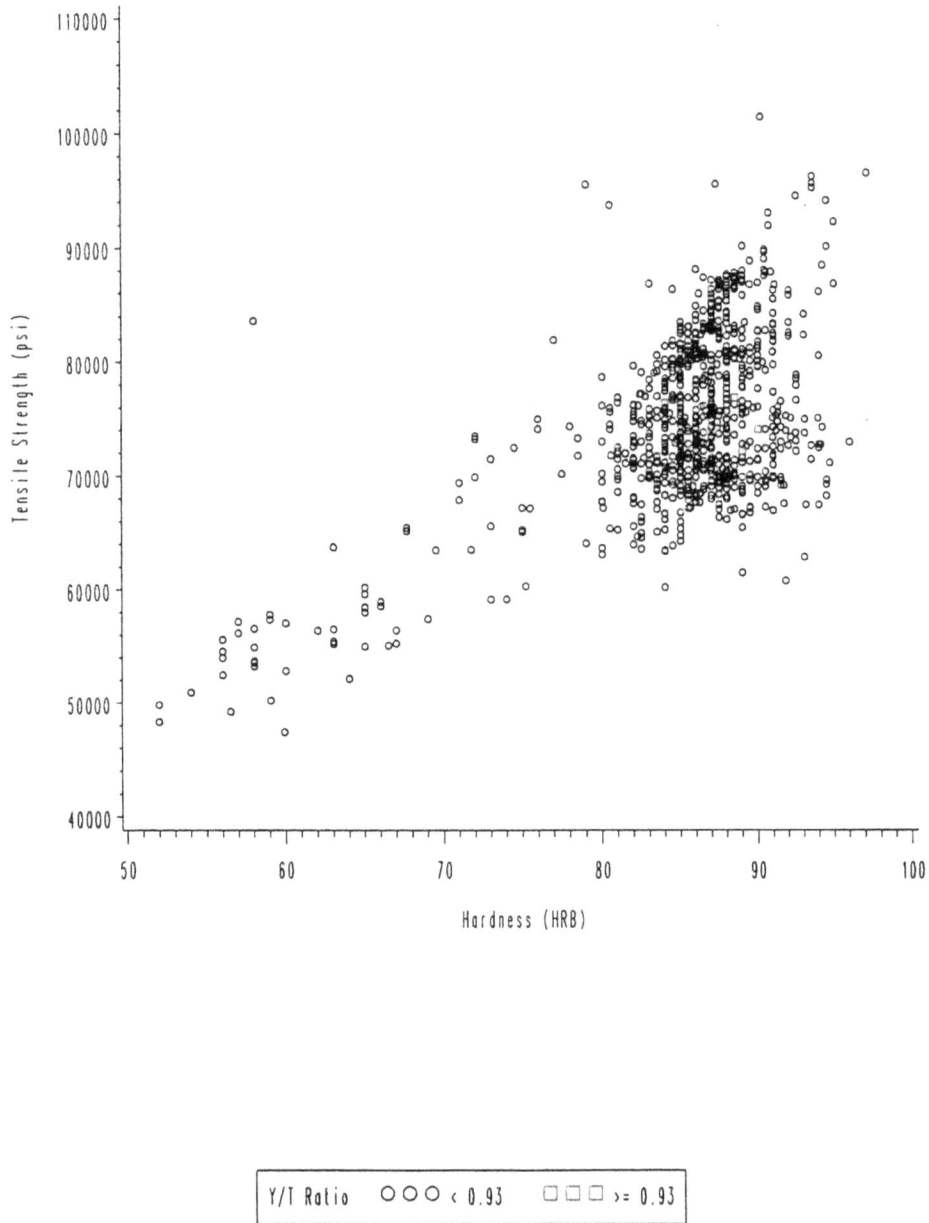

Figure 3.6. Ultimate tensile strength versus Rockwell B hardness by yield-to-tensile ratio category (distinguished by unique symbol types) — collected data with tensile and hardness property measures.

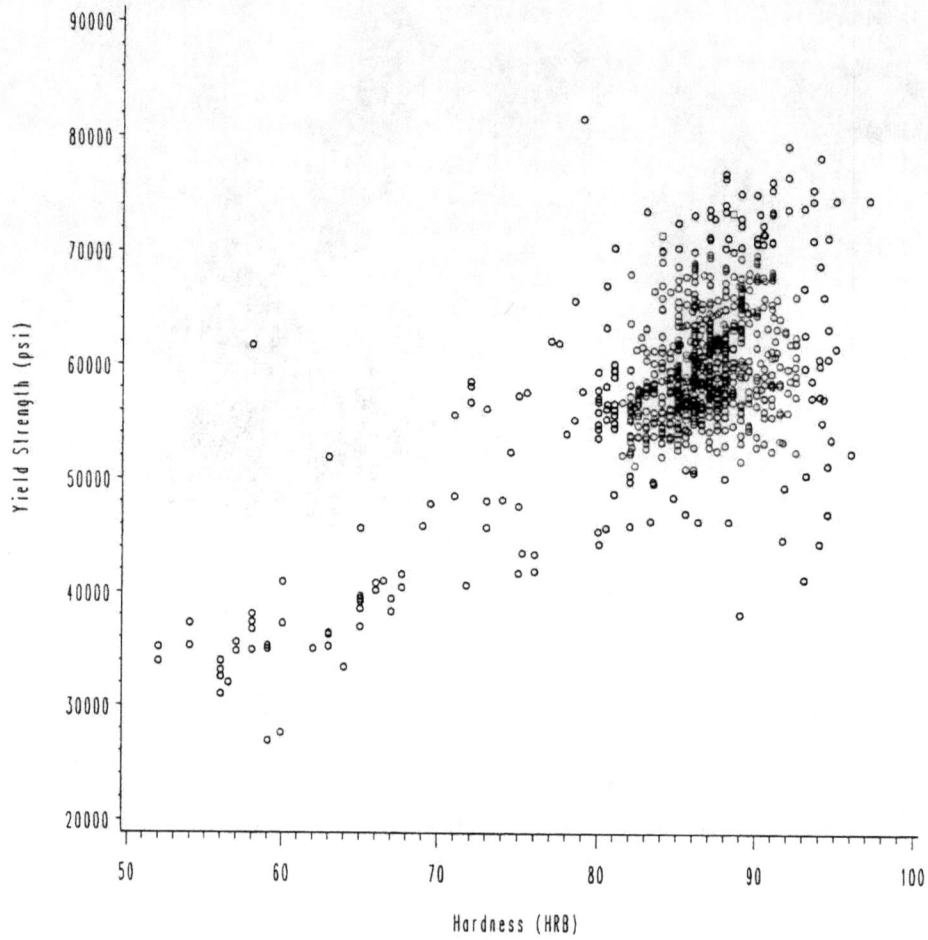

Figure 3.7. **Yield strength versus Rockwell B hardness by yield-to-tensile ratio category (distinguished by unique symbol types) — collected data with tensile and hardness property measures.**

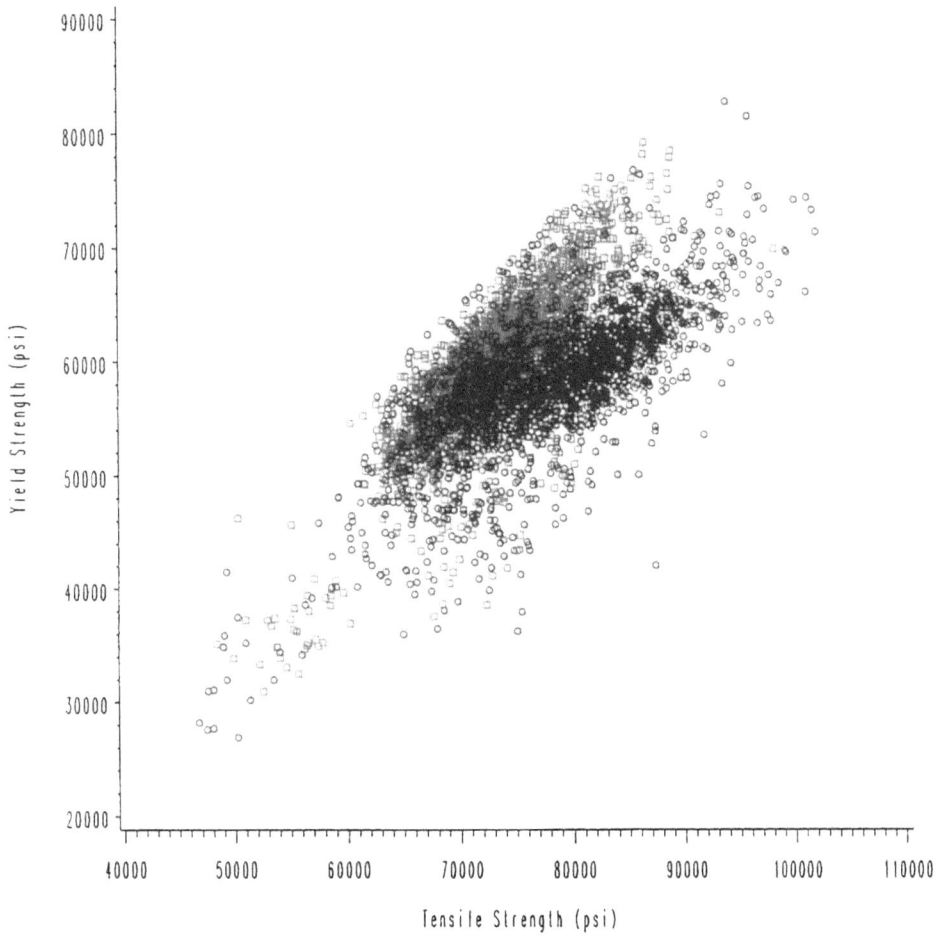

Figure 3.8. Yield strength versus ultimate tensile strength by pipe type (distinguished by unique symbol types) — collected data with tensile property measures.

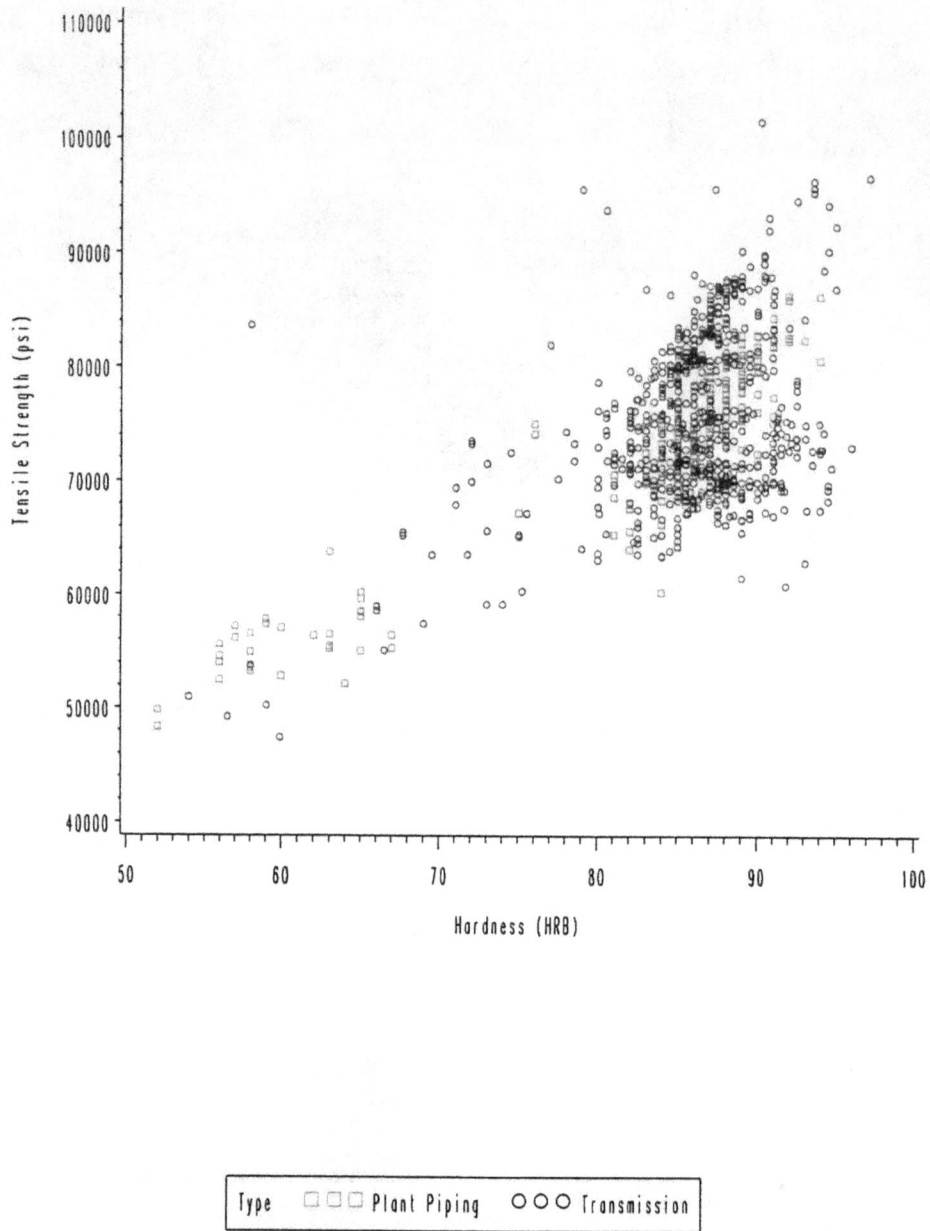

Figure 3.9. Ultimate tensile strength versus Rockwell B hardness by pipe type (distinguished by unique symbol types) — collected data with tensile and hardness property measures.

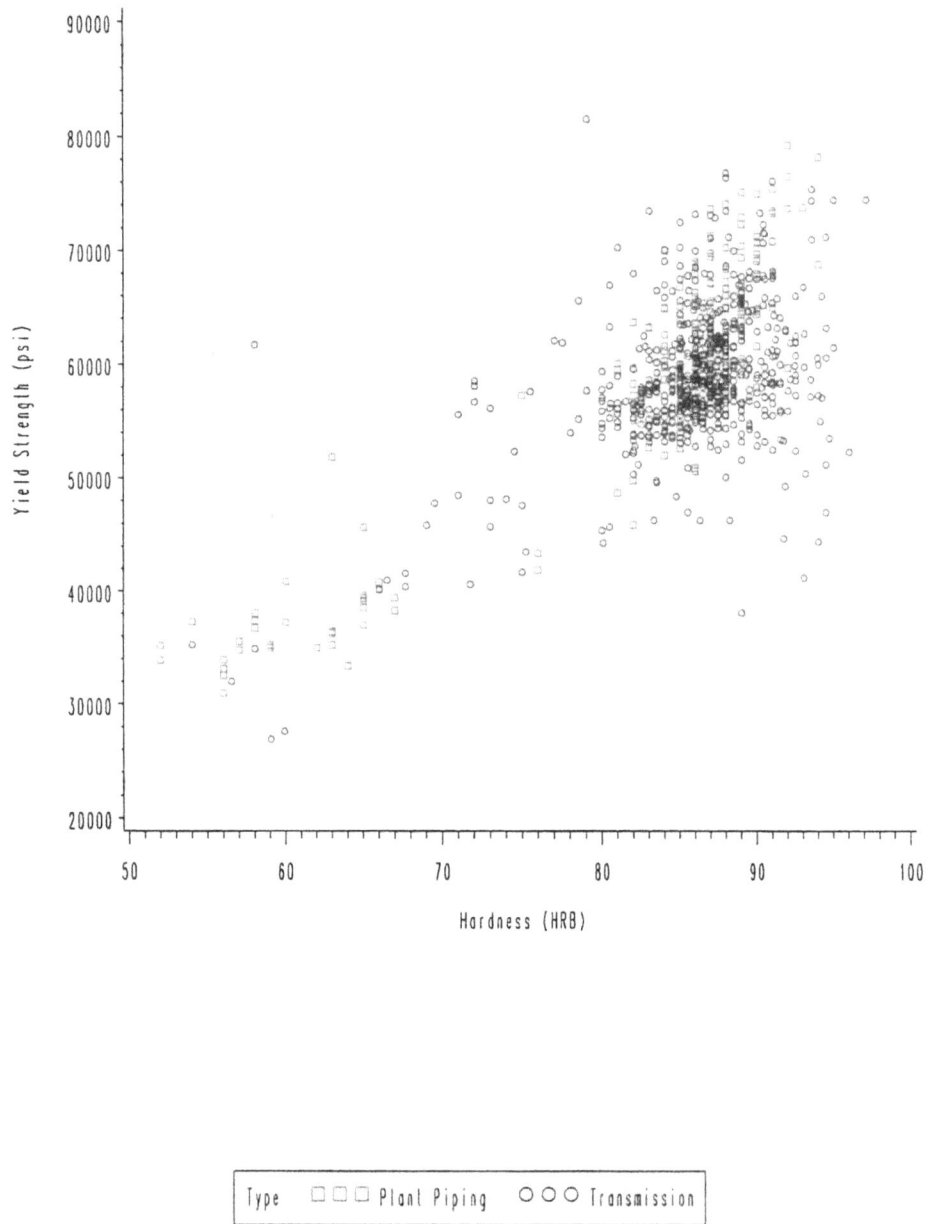

Figure 3.10. Yield strength versus Rockwell B hardness by pipe type (distinguished by unique symbol types) — collected data with tensile and hardness property measures.

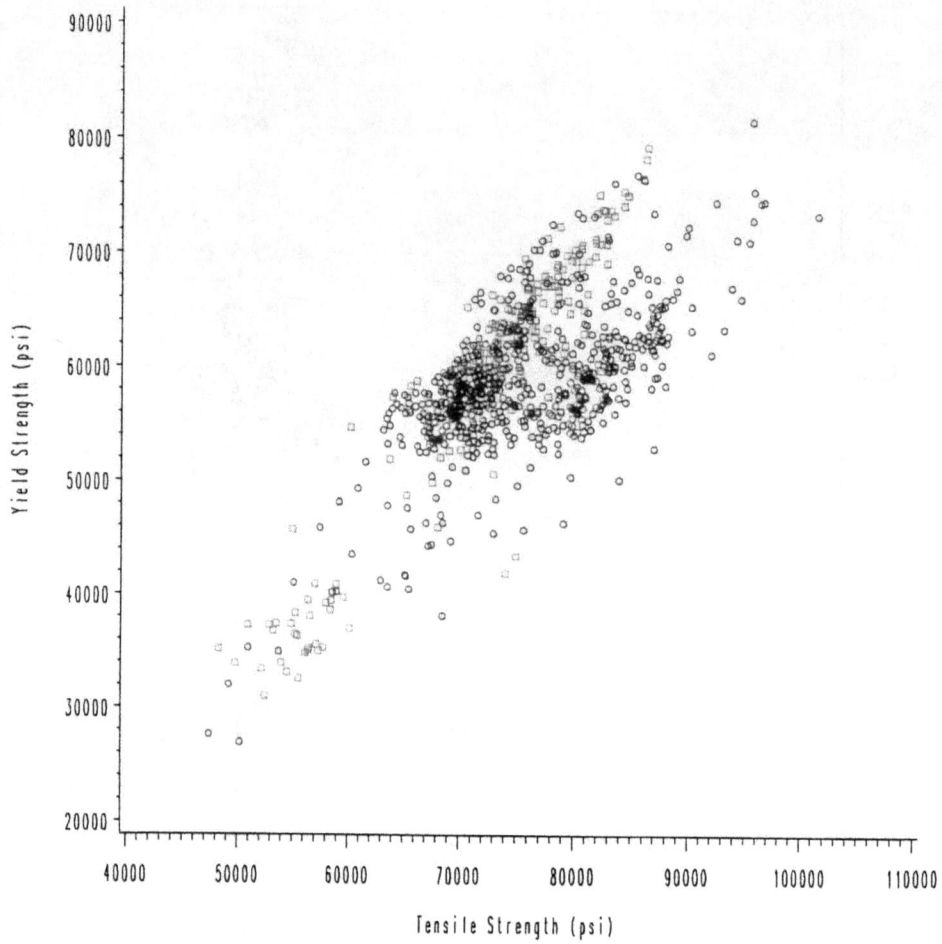

Figure 3.11. Yield strength versus ultimate tensile strength by pipe type (distinguished by unique symbol types) — collected data with tensile and hardness property measures

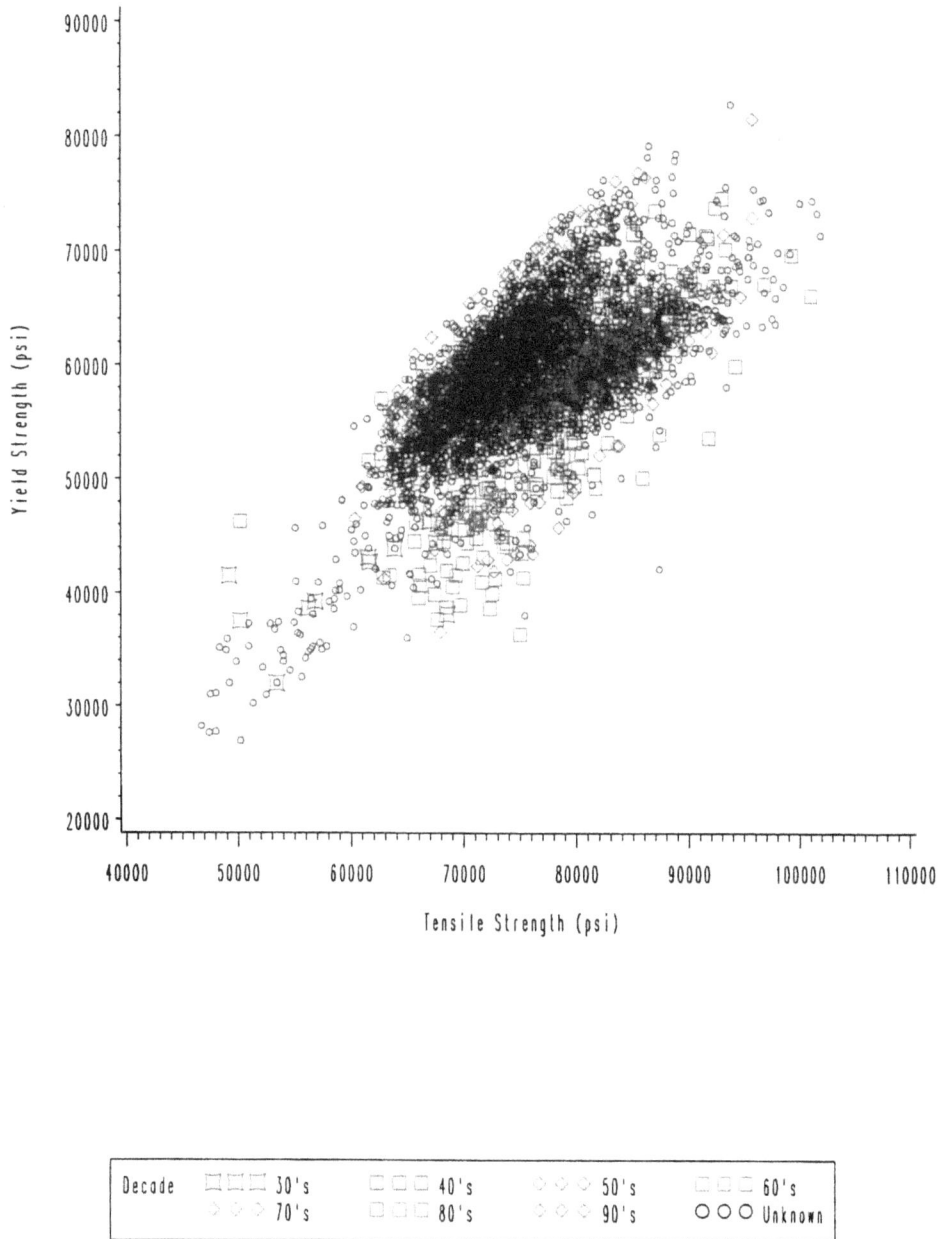

Figure 3.12. Yield strength versus ultimate tensile strength by decade of manufacture (distinguished by unique symbol types) — collected data with tensile property measures.

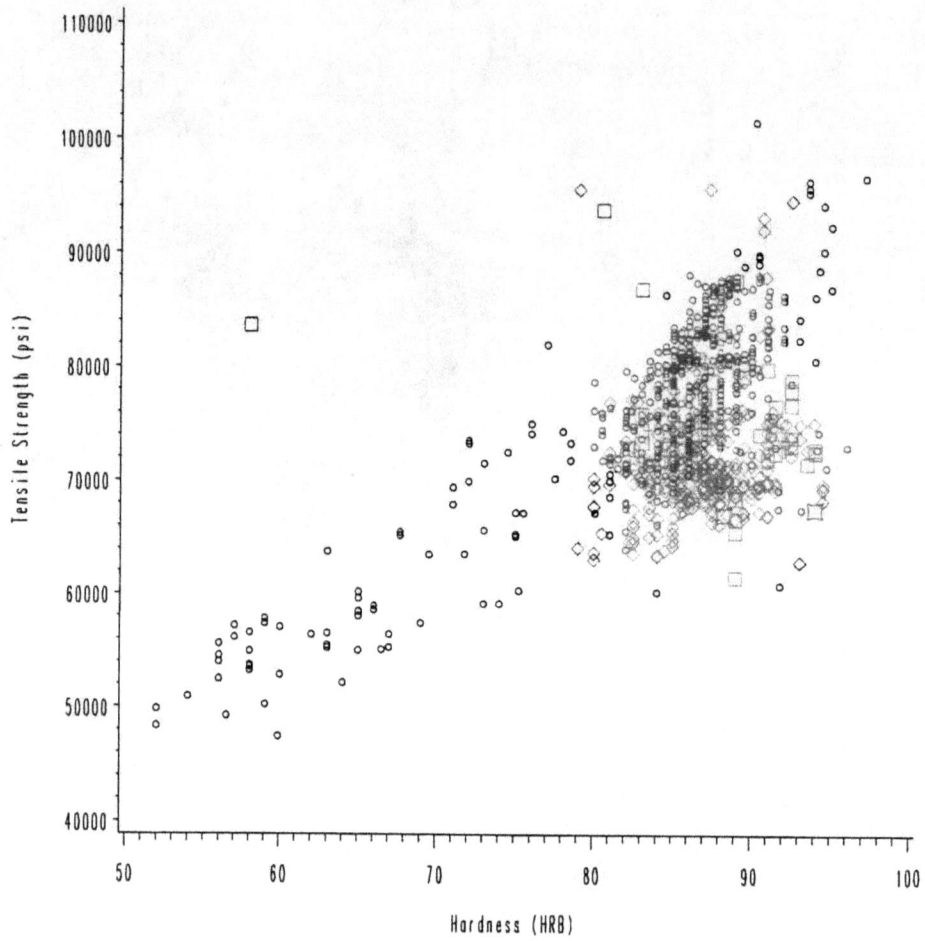

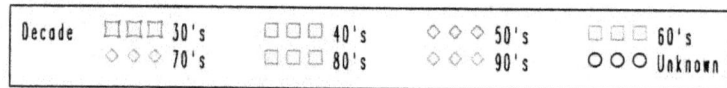

Figure 3.13. Ultimate tensile strength versus Rockwell B hardness by decade of manufacture (distinguished by unique symbol types) — collected data with tensile and hardness property measures

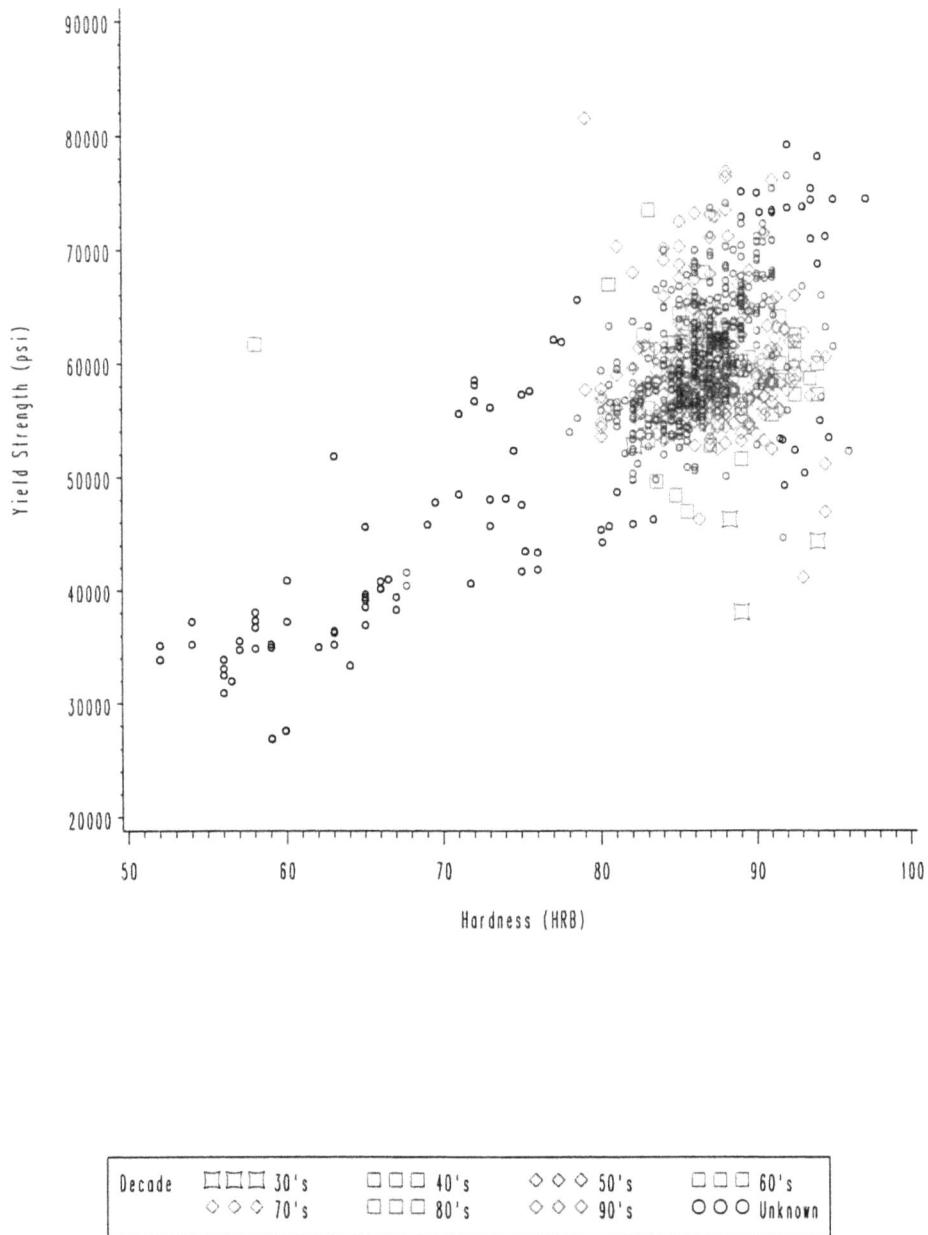

Figure 3.14. Yield strength versus Rockwell B hardness by decade of manufacture (distinguished by unique symbol types) — collected data with tensile and hardness property measures

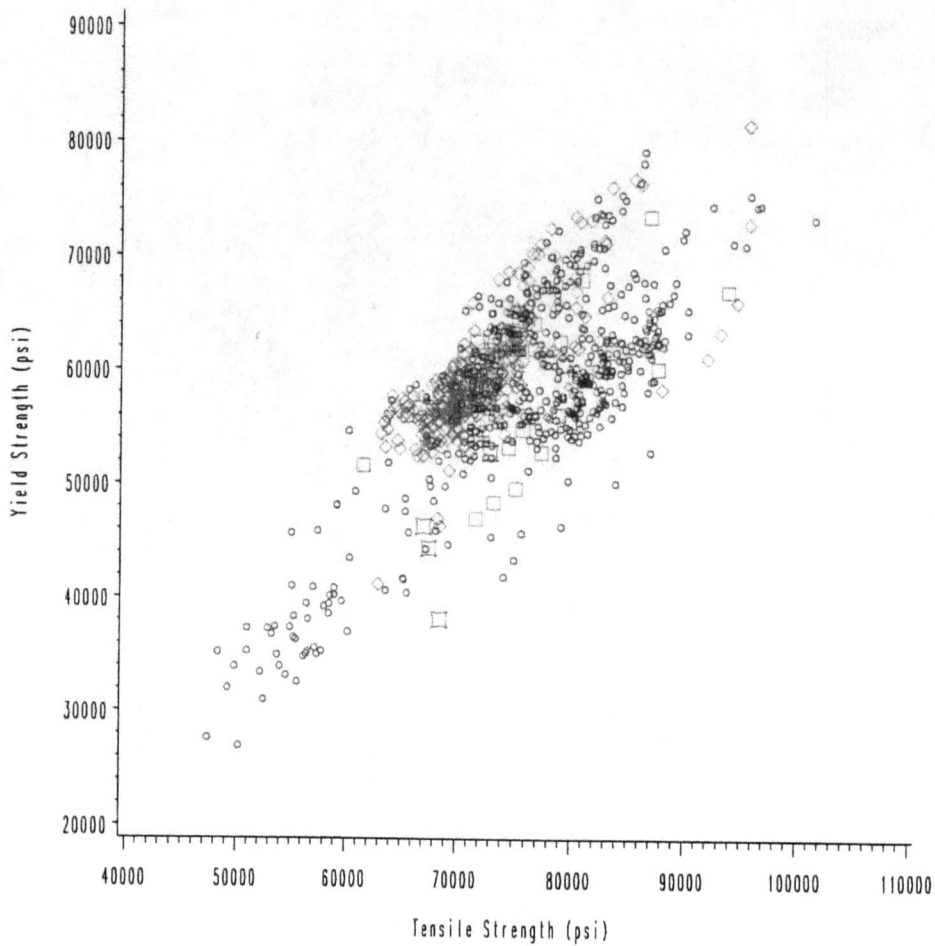

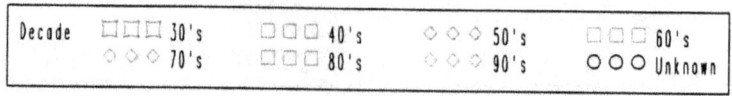

Figure 3.15. Yield strength versus ultimate tensile strength by decade of manufacture (distinguished by unique symbol types) — collected data with tensile and hardness property measures

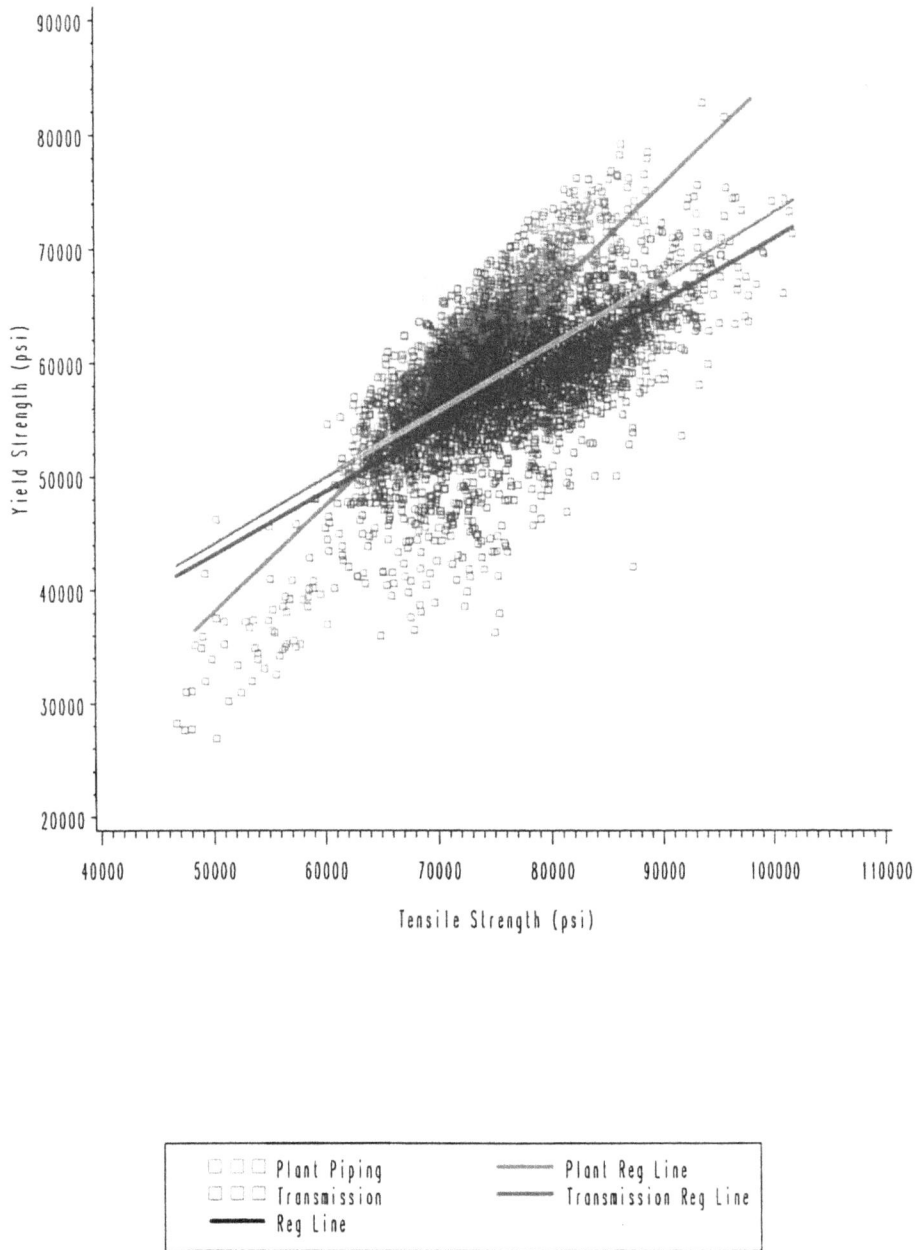

Figure 3.16. Yield strength versus ultimate tensile strength by pipe type with superimposed linear regression models (distinguished by unique symbol types) fitted by pipe type — collected data with tensile property measures

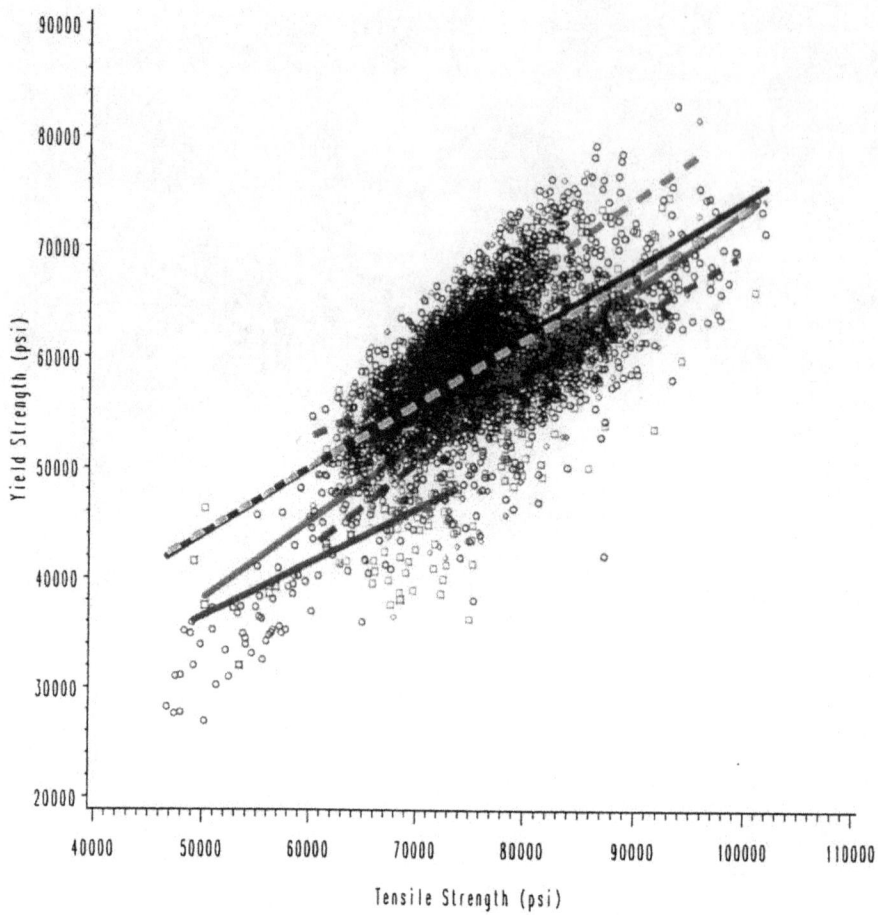

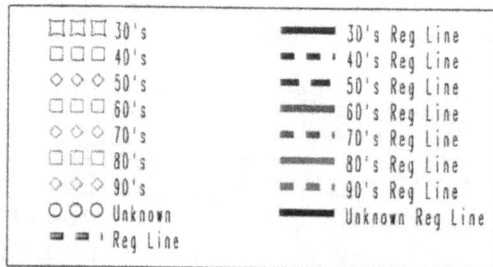

Figure 3.17. Yield strength versus ultimate tensile strength by decade of manufacture with superimposed linear regression models (distinguished by unique symbol types) fitted by decade of manufacture — collected data with tensile property measures

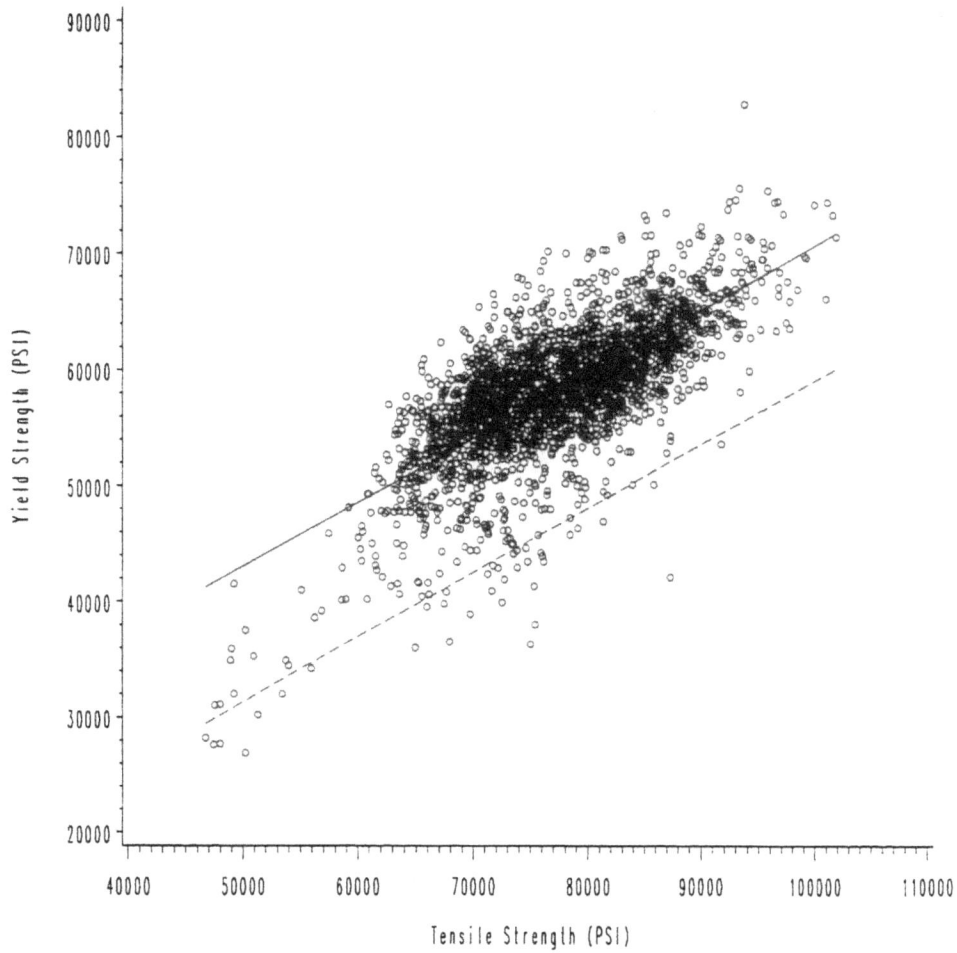

Yield Strength = 9,699 + 0.06805 * Tensile Strength

Figure 3.18. Yield strength versus ultimate tensile strength with estimated linear model of yield strength (solid line) and associated 99 percent lower confidence bound on 0.5 percentile yield strength (dashed line) — collected data with tensile property measures for transmission line lengths manufactured prior to 1980

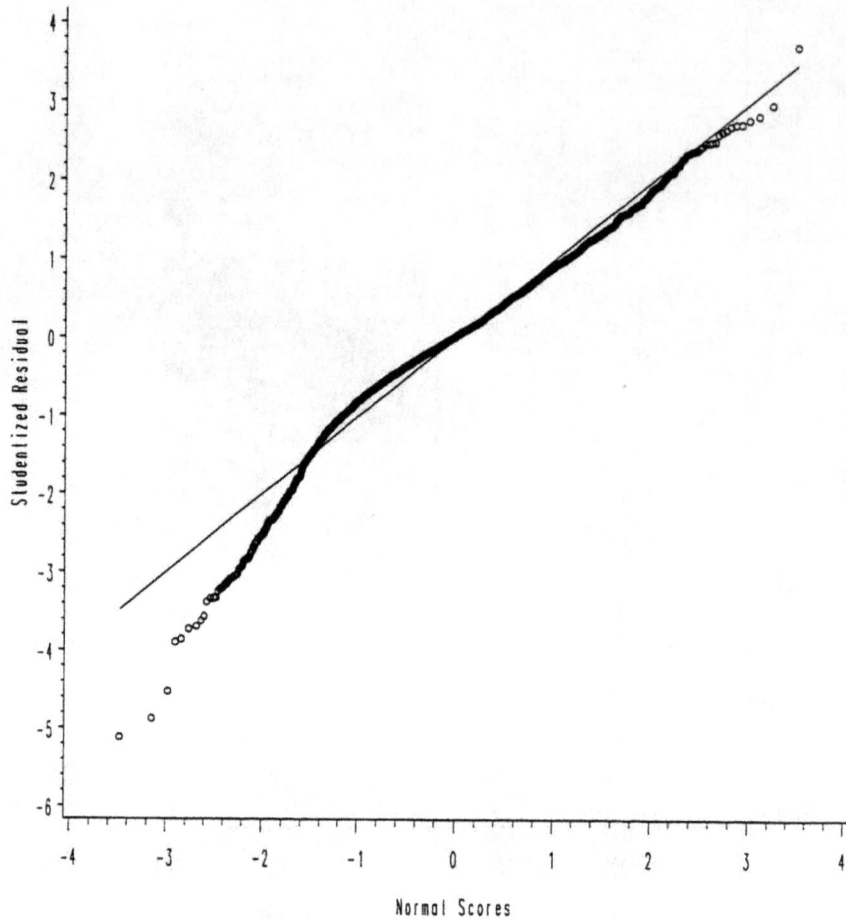

Figure 3.19. Normal probability plot for scatter about estimated yield strength versus ultimate tensile strength relationship — collected data with tensile property measures for transmission line lengths manufactured prior to 1980

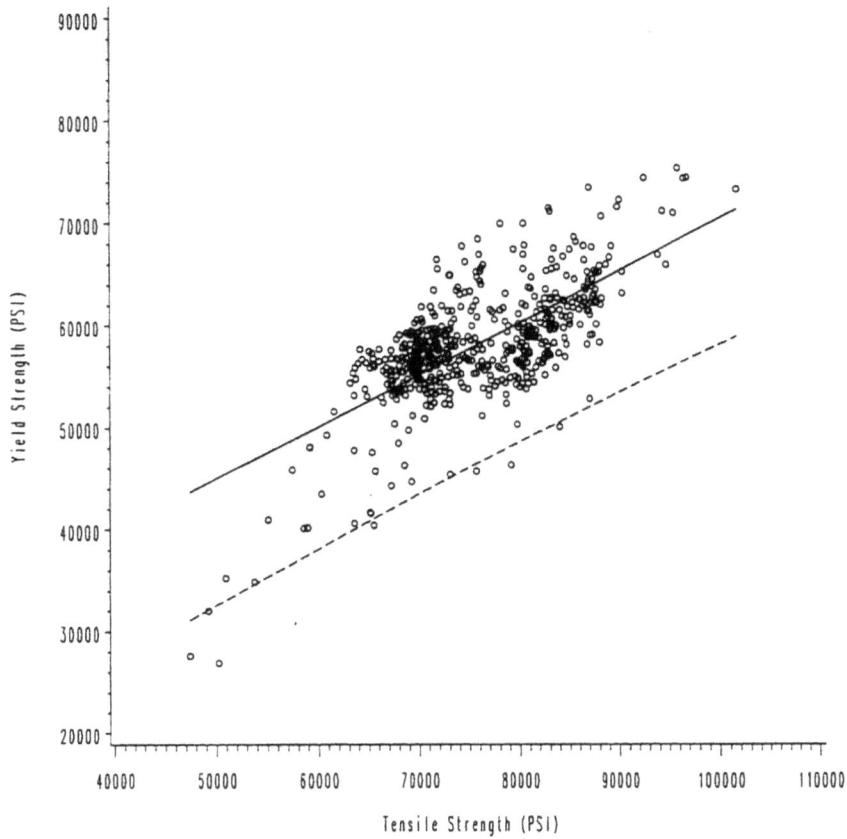

Figure 3.20. Yield strength versus ultimate tensile strength with estimated linear model of yield strength (solid line) and associated 99 percent lower confidence bound on 0.5 percentile yield strength (dashed line) — collected data with tensile and hardness property measures for transmission line lengths manufactured prior to 1980

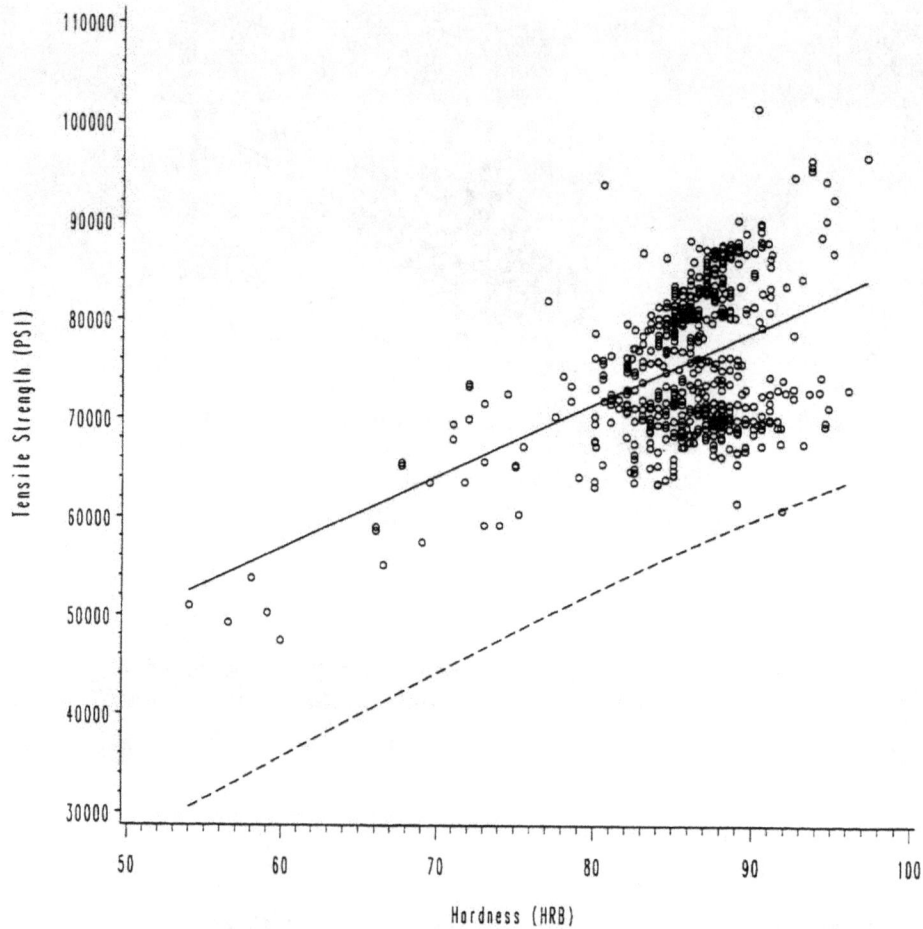

Tensile Strength = 12,948 + 731.64289 * Hardness

Figure 3.21. **Ultimate tensile strength versus Rockwell B hardness with estimated linear model of tensile strength (solid line) and associated 99 percent lower confidence bound on the 0.5 percentile tensile strength (dashed line) — collected data with tensile and hardness property measures for transmission line lengths manufactured prior to 1980**

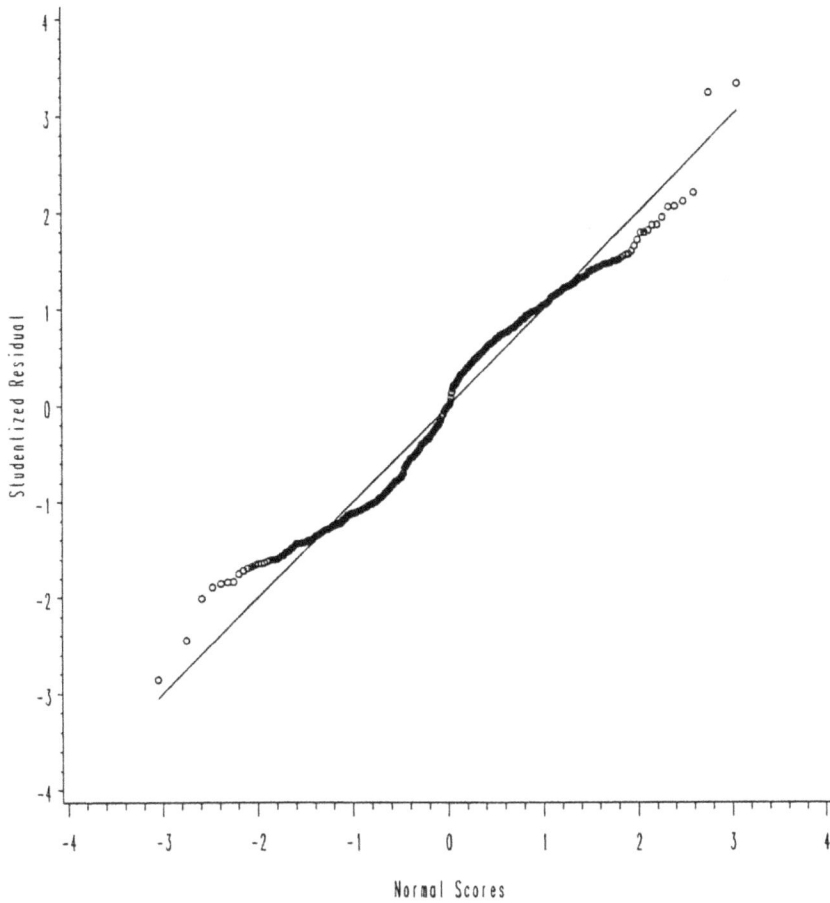

Figure 3.22. Normal probability plot for scatter about estimated ultimate tensile strength versus Rockwell B hardness relationship — collected data with tensile and hardness property measures for transmission line lengths manufactured prior to 1980

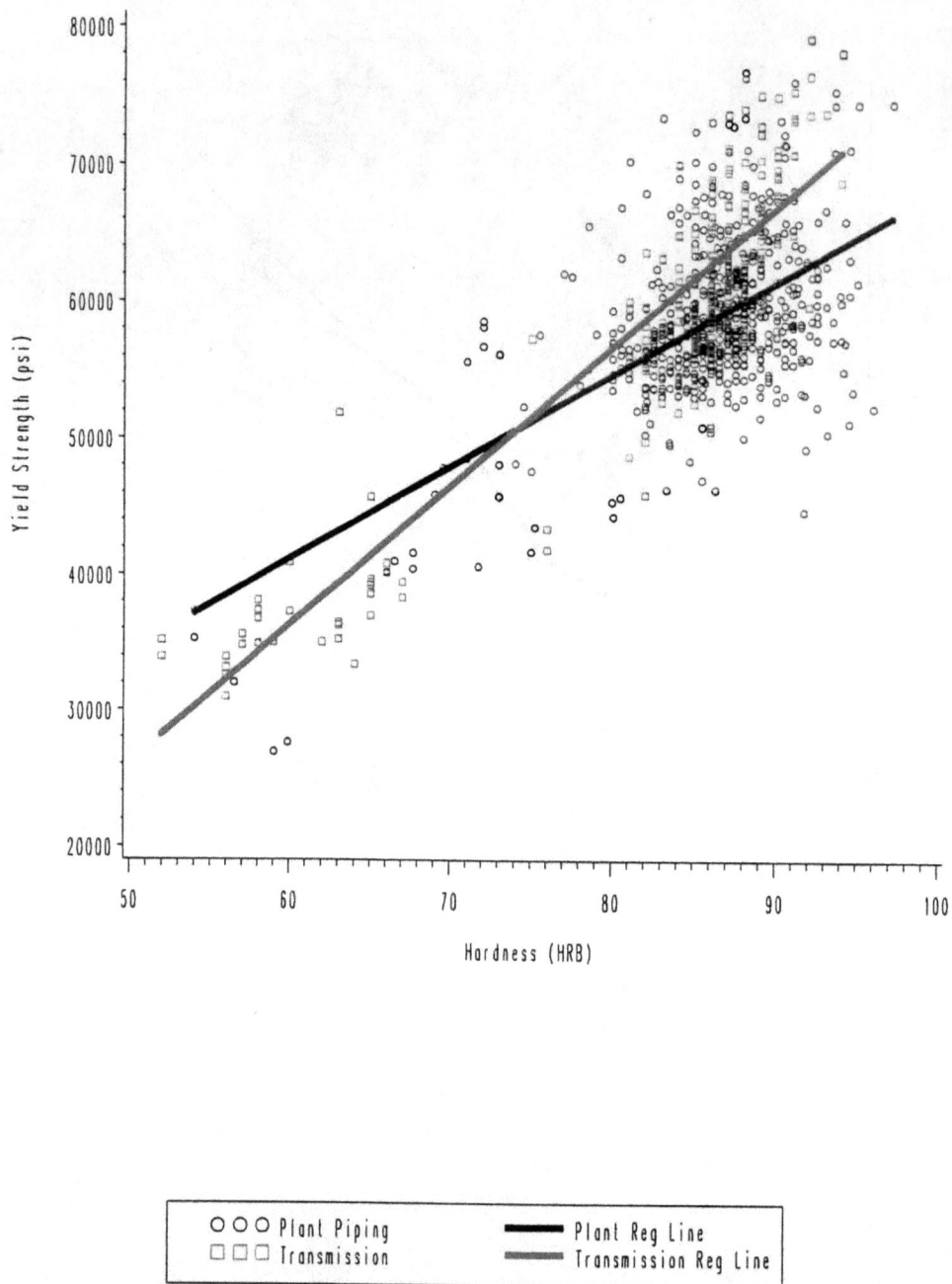

Figure 3.23. Yield strength versus Rockwell B hardness by pipe type with superimposed linear regression models (distinguished by unique line types) fitted by pipe type — collected data with tensile and hardness property measures

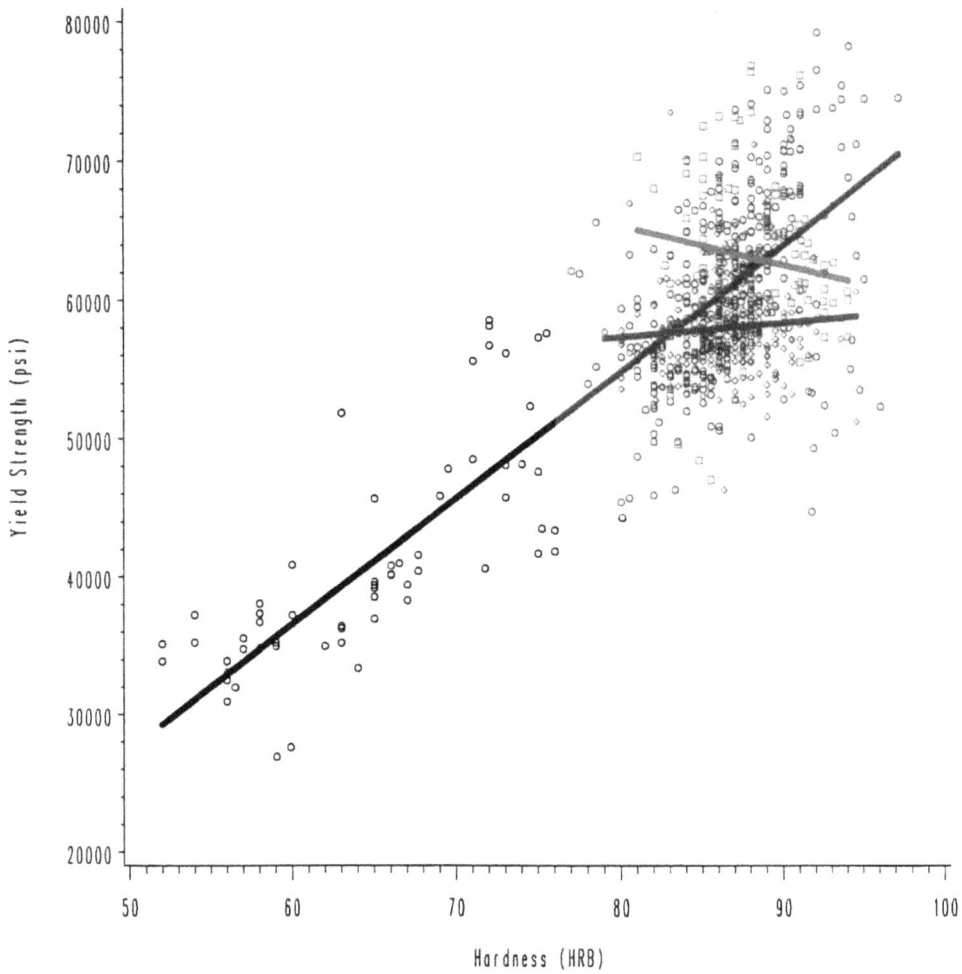

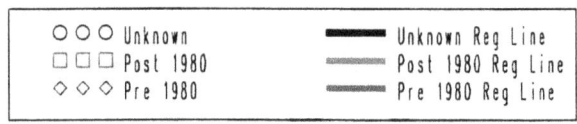

Figure 3.24. Yield strength versus Rockwell B hardness by manufacture date interval with superimposed linear regression models (distinguished by unique line types) fitted by manufacture date interval — collected data with tensile and hardness property measures

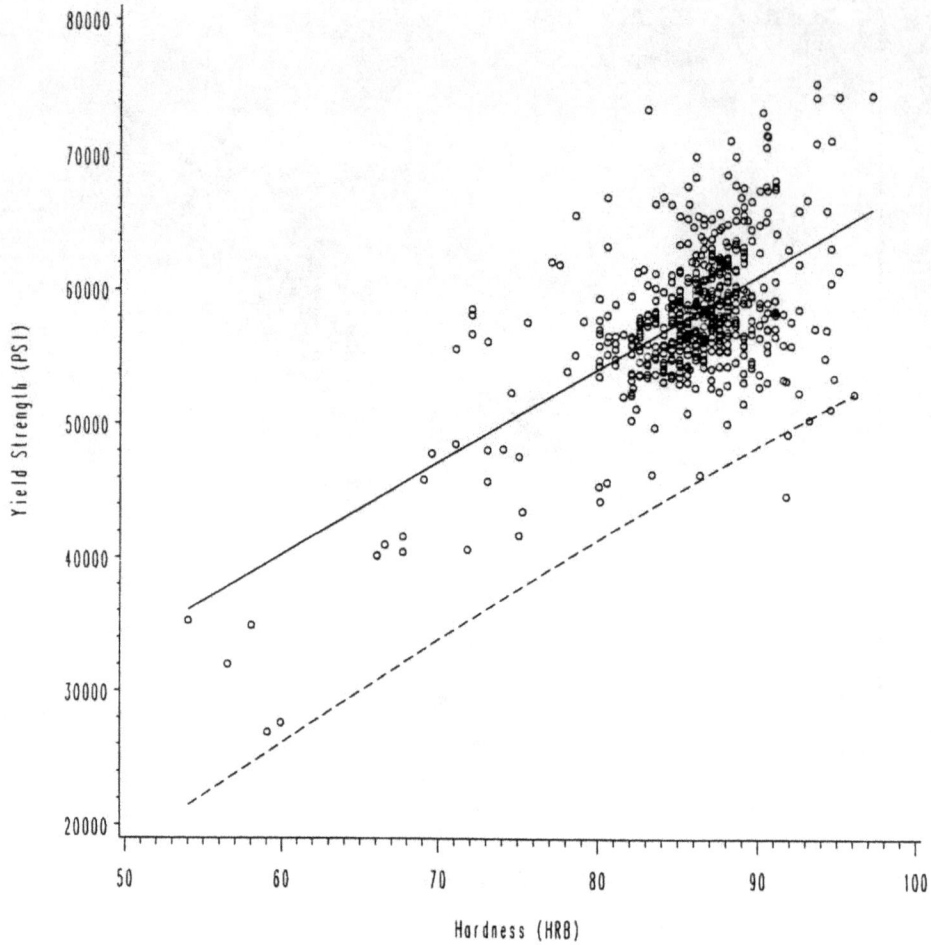

Yield Strength = -1,425 + 694.85075 * Hardness

Figure 3.25. Yield strength versus Rockwell B hardness with estimated linear model of yield strength (solid line) and associated 99 percent lower confidence bound on the 0.5 percentile yield strength (dashed line) — collected data with tensile and hardness property measures for transmission line lengths manufactured prior to 1980

70

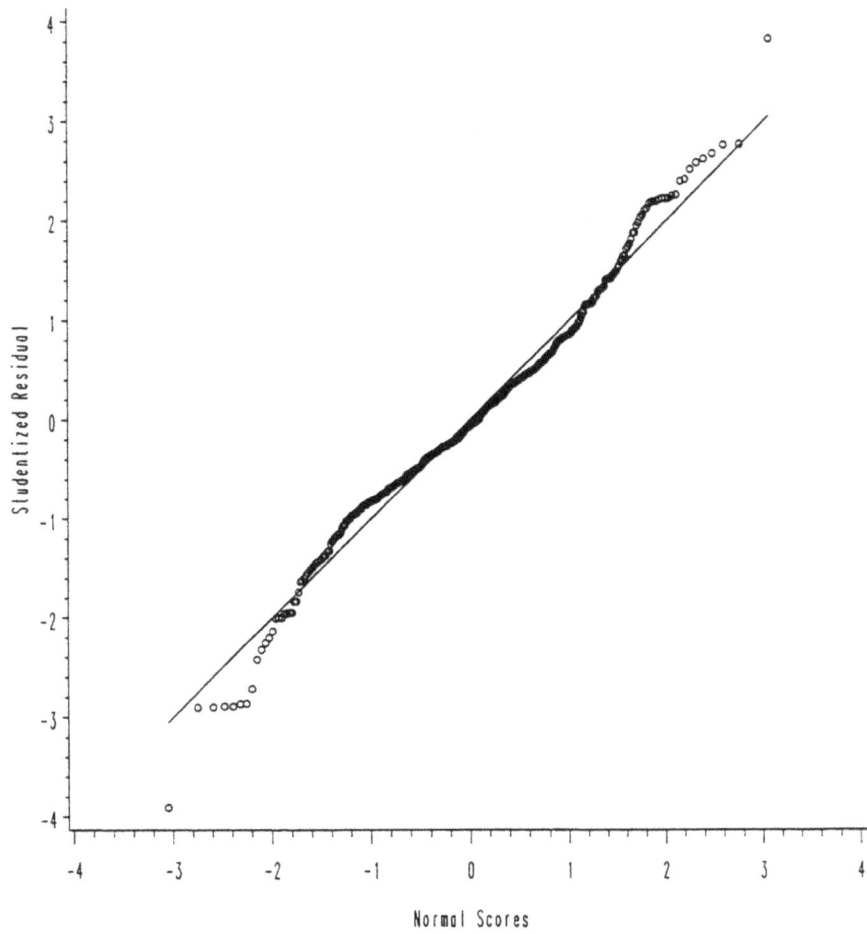

Figure 3.26. Normal probability plot for scatter about estimated yield strength versus Rockwell B hardness relationship — collected data with tensile and hardness property measures for transmission line lengths manufactured prior to 1980

71

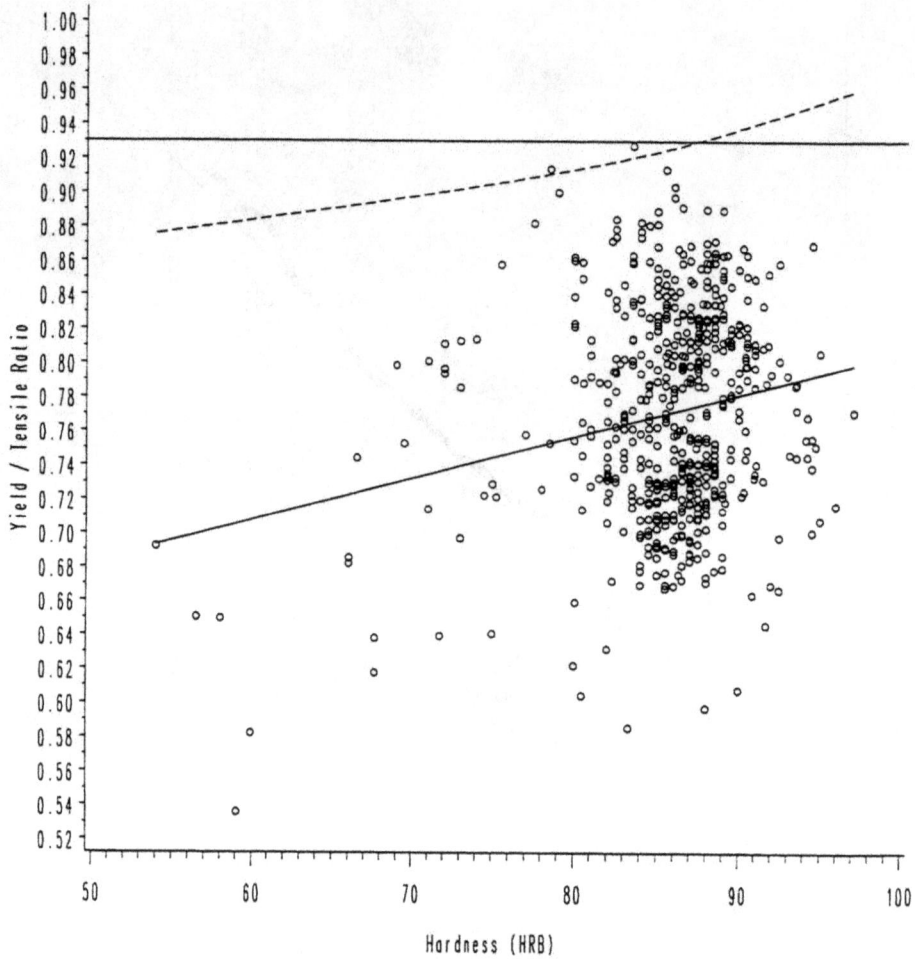

Figure 3.27. Yield-to-tensile strength ratio versus Rockwell B hardness with estimated linear model of yield-to-tensile strength ratio (solid line) and associated 99 percent upper confidence bound on the 99 percentile yield-to-tensile strength ratio (dashed line) — collected data with tensile and hardness property measures for transmission line lengths manufactured prior to 1980

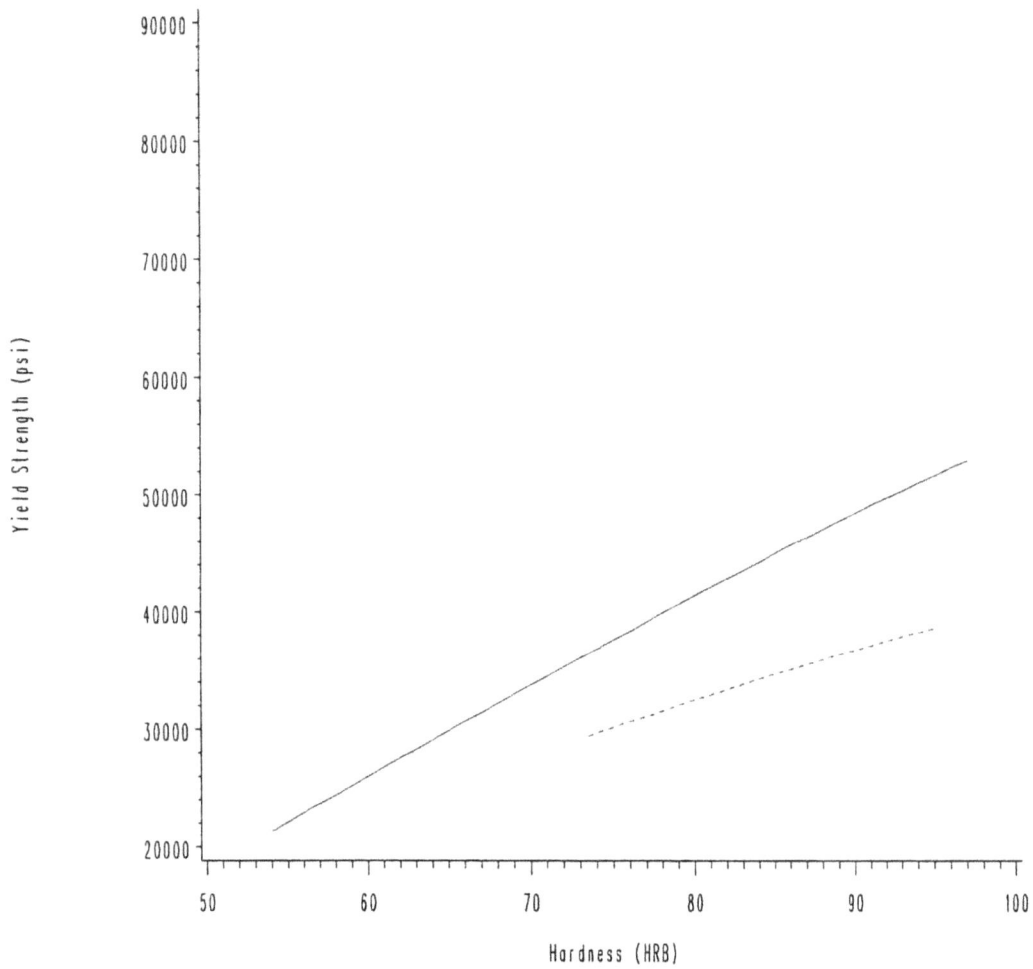

Figure 3.28. Developed relationship between yield strength and Rockwell B hardness as established directly (solid line) and indirectly through ultimate tensile strength (dashed line)

Yield Strength (psi)		FREQ.	CUM. FREQ.	PCT.	CUM. PCT.
40000		0	0	0.00	0.00
42000		0	0	0.00	0.00
44000		0	0	0.00	0.00
46000		0	0	0.00	0.00
48000		0	0	0.00	0.00
50000		2	2	0.51	0.51
52000		1	3	0.25	0.76
54000		9	12	2.28	3.05
56000		32	44	8.12	11.17
58000		48	92	12.18	23.35
60000		76	168	19.29	42.64
62000		82	250	20.81	63.45
64000		74	324	18.78	82.23
66000		43	367	10.91	93.15
68000		19	386	4.82	97.97
70000		2	388	0.51	98.48
72000		3	391	0.76	99.24
74000		1	392	0.25	99.49
76000		0	392	0.00	99.49
78000		2	394	0.51	100.00
80000		0	394	0.00	100.00

FREQUENCY

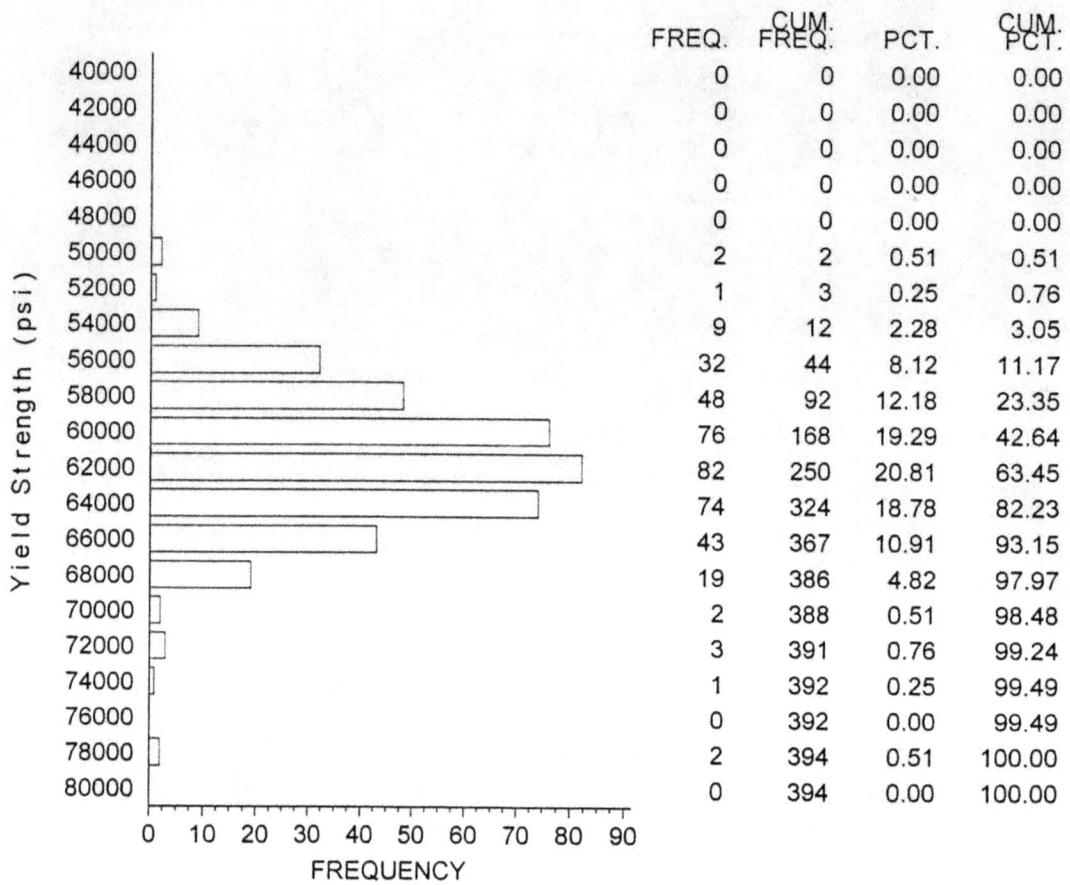

Figure 3.29. Frequency histogram and associated percentiles of yield strengths for heats of X52 grade pipe lengths

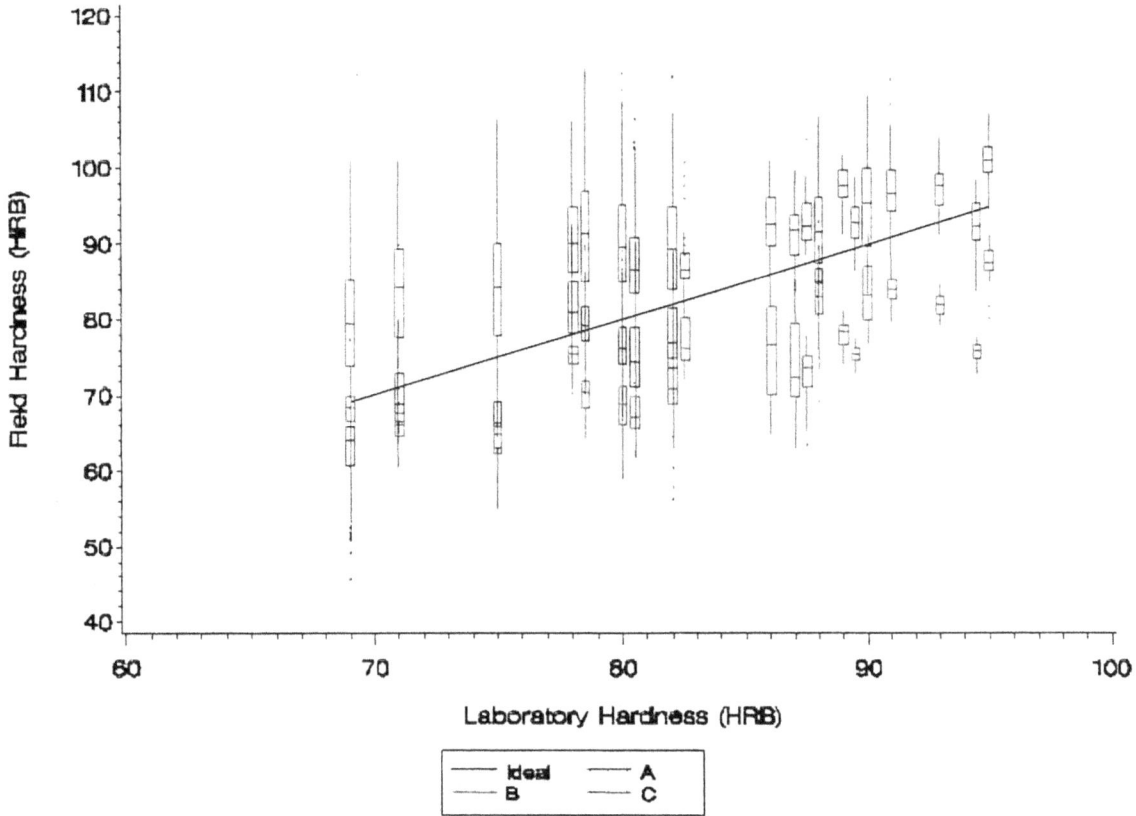

Figure 4.1 Comparison of the results from 3 portable hardness testers that were evaluated for determining "field" hardness.

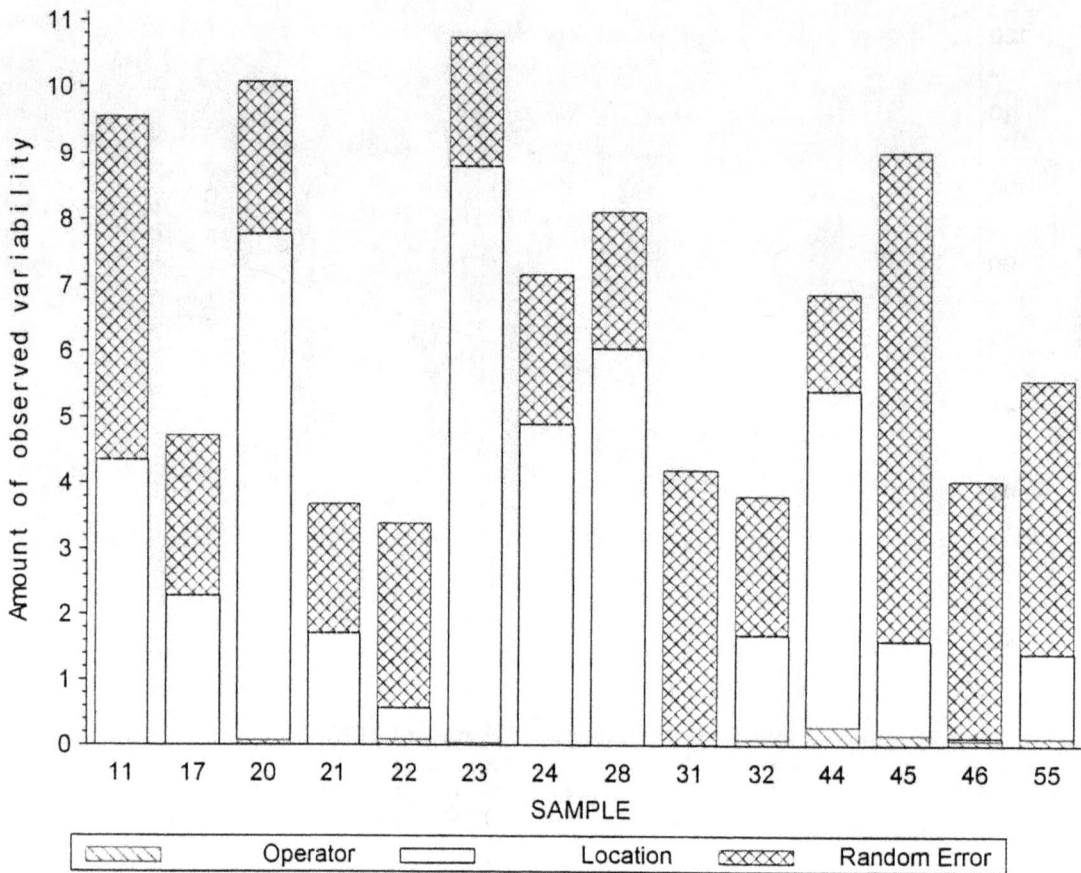

Figure 4.2. Percentile of overall variability in Equotip[©] measured field hardness attributable to operator, location on length, and random unknown factors based on collected Equotip[©] field hardness measurement characterization data

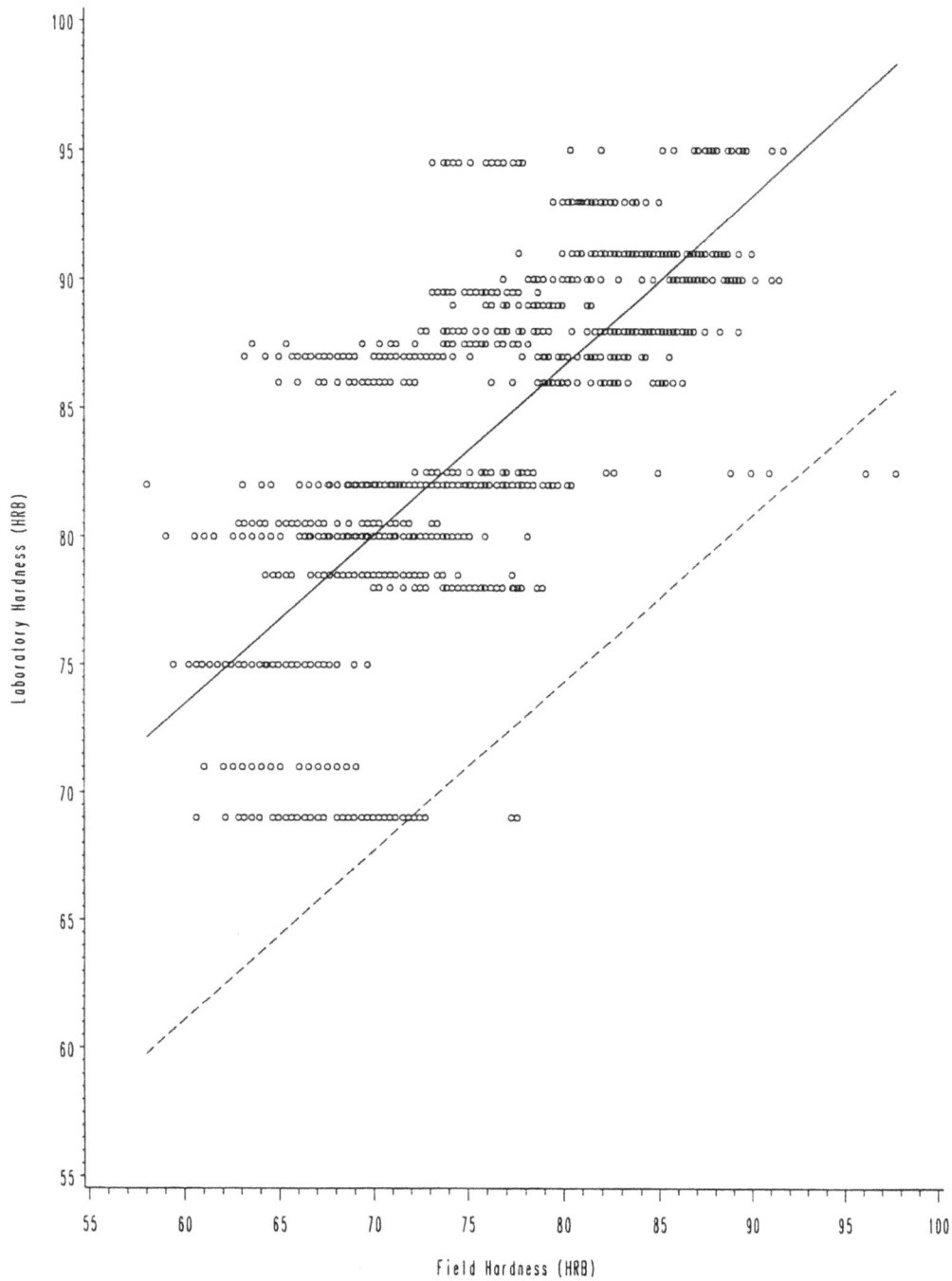

Figure 4.3. Rockwell B hardness as measured using laboratory techniques versus hardness as measured using Equotip[C] tester, with estimated linear model (solid line) and 99 percent confidence bound on the 0.5 percentile laboratory hardness (dashed line) — collected Equotip[C] field hardness measurement characterization data

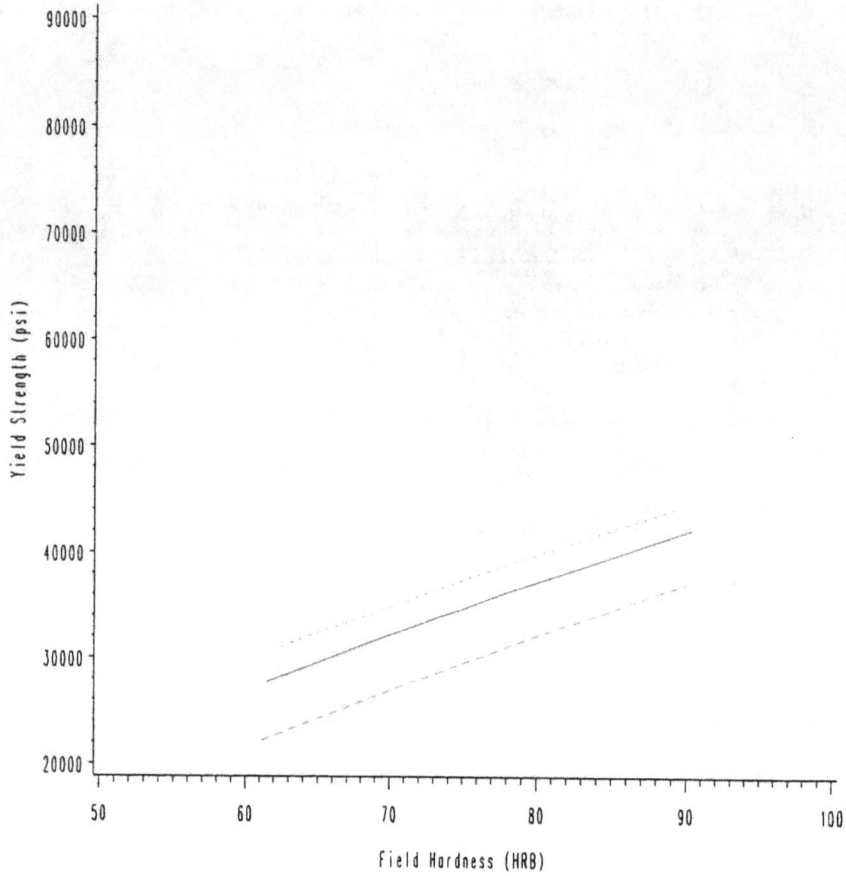

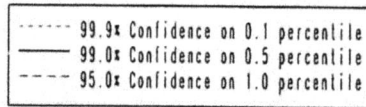

Figure 4.4. Developed relationship between yield strength and Equotip© measured field hardness as a function of confidence level and percentile (distinguished by unique line type) based on collected Equotip© field hardness measurement characterization data

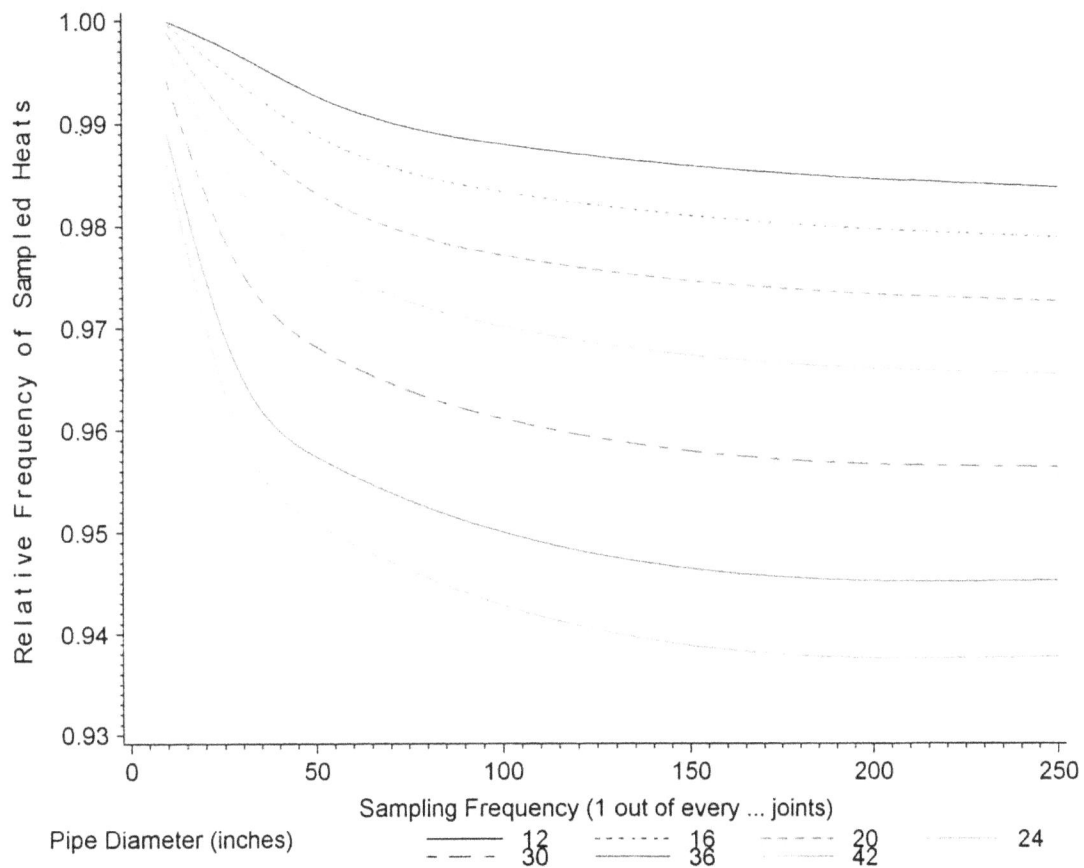

Figure 5.1. Relative frequency of sampled heats (within a pipeline) versus length sampling frequency by pipe diameter (distinguished by unique line type) developed from simulation analyses

APPENDIX A
DETAILED DATA TABLES

Table A.3.3. Estimated lower tolerance bound yield strength as function of measured ultimate tensile strength, confidence level, and percentile — based on collected data with tensile property measures for transmission line joints manufactured prior to 1980

Targeted Percentile	Confidence Level	Measured Ultimate Tensile Strength (PSI)	Estimated Lower Tolerance Bound on Yield Strength (PSI)
0.1%	99.9%	47,000	26,700
0.1%	99.9%	48,000	27,300
0.1%	99.9%	49,000	27,900
0.1%	99.9%	50,000	28,500
0.1%	99.9%	51,000	29,000
0.1%	99.9%	52,000	29,600
0.1%	99.9%	53,000	30,200
0.1%	99.9%	54,000	30,800
0.1%	99.9%	55,000	31,300
0.1%	99.9%	56,000	31,900
0.1%	99.9%	57,000	32,500
0.1%	99.9%	58,000	33,100
0.1%	99.9%	59,000	33,600
0.1%	99.9%	60,000	34,200
0.1%	99.9%	61,000	34,800
0.1%	99.9%	62,000	35,300
0.1%	99.9%	63,000	35,900
0.1%	99.9%	64,000	36,500
0.1%	99.9%	65,000	37,000
0.1%	99.9%	66,000	37,600
0.1%	99.9%	67,000	38,100
0.1%	99.9%	68,000	38,700
0.1%	99.9%	69,000	39,300
0.1%	99.9%	70,000	39,800
0.1%	99.9%	71,000	40,400
0.1%	99.9%	72,000	41,000
0.1%	99.9%	73,000	41,500
0.1%	99.9%	74,000	42,100
0.1%	99.9%	75,000	42,600
0.1%	99.9%	76,000	43,200
0.1%	99.9%	77,000	43,700
0.1%	99.9%	78,000	44,300
0.1%	99.9%	79,000	44,800
0.1%	99.9%	80,000	45,400
0.1%	99.9%	81,000	45,900
0.1%	99.9%	82,000	46,500
0.1%	99.9%	83,000	47,000
0.1%	99.9%	84,000	47,600
0.1%	99.9%	85,000	48,100
0.1%	99.9%	86,000	48,700
0.1%	99.9%	87,000	49,200
0.1%	99.9%	88,000	49,700
0.1%	99.9%	89,000	50,300
0.1%	99.9%	90,000	50,800
0.1%	99.9%	91,000	51,400
0.1%	99.9%	92,000	51,900
0.1%	99.9%	93,000	52,400
0.1%	99.9%	94,000	53,000
0.1%	99.9%	95,000	53,500
0.1%	99.9%	96,000	54,100
0.1%	99.9%	97,000	54,600
0.1%	99.9%	98,000	55,100
0.1%	99.9%	99,000	55,700
0.1%	99.9%	100,000	56,200

Table A.3.3. Estimated lower tolerance bound yield strength as function of measured ultimate tensile strength, confidence level, and percentile — based on collected data with tensile property measures for transmission line joints manufactured prior to 1980 (Continued)

Targeted Percentile	Confidence Level	Measured Ultimate Tensile Strength (PSI)	Estimated Lower Tolerance Bound on Yield Strength (PSI)
0.5%	99.9%	47,000	29,100
0.5%	99.9%	48,000	29,600
0.5%	99.9%	49,000	30,200
0.5%	99.9%	50,000	30,800
0.5%	99.9%	51,000	31,400
0.5%	99.9%	52,000	31,900
0.5%	99.9%	53,000	32,500
0.5%	99.9%	54,000	33,100
0.5%	99.9%	55,000	33,700
0.5%	99.9%	56,000	34,200
0.5%	99.9%	57,000	34,800
0.5%	99.9%	58,000	35,400
0.5%	99.9%	59,000	36,000
0.5%	99.9%	60,000	36,500
0.5%	99.9%	61,000	37,100
0.5%	99.9%	62,000	37,700
0.5%	99.9%	63,000	38,200
0.5%	99.9%	64,000	38,800
0.5%	99.9%	65,000	39,400
0.5%	99.9%	66,000	39,900
0.5%	99.9%	67,000	40,500
0.5%	99.9%	68,000	41,100
0.5%	99.9%	69,000	41,600
0.5%	99.9%	70,000	42,200
0.5%	99.9%	71,000	42,800
0.5%	99.9%	72,000	43,300
0.5%	99.9%	73,000	43,900
0.5%	99.9%	74,000	44,400
0.5%	99.9%	75,000	45,000
0.5%	99.9%	76,000	45,500
0.5%	99.9%	77,000	46,100
0.5%	99.9%	78,000	46,600
0.5%	99.9%	79,000	47,200
0.5%	99.9%	80,000	47,700
0.5%	99.9%	81,000	48,300
0.5%	99.9%	82,000	48,800
0.5%	99.9%	83,000	49,400
0.5%	99.9%	84,000	49,900
0.5%	99.9%	85,000	50,500
0.5%	99.9%	86,000	51,000
0.5%	99.9%	87,000	51,600
0.5%	99.9%	88,000	52,100
0.5%	99.9%	89,000	52,600
0.5%	99.9%	90,000	53,200
0.5%	99.9%	91,000	53,700
0.5%	99.9%	92,000	54,300
0.5%	99.9%	93,000	54,800
0.5%	99.9%	94,000	55,300
0.5%	99.9%	95,000	55,900
0.5%	99.9%	96,000	56,400
0.5%	99.9%	97,000	56,900
0.5%	99.9%	98,000	57,500
0.5%	99.9%	99,000	58,000
0.5%	99.9%	100,000	58,500

Table A.3.3. Estimated lower tolerance bound yield strength as function of measured ultimate tensile strength, confidence level, and percentile — based on collected data with tensile property measures for transmission line joints manufactured prior to 1980 (Continued)

Targeted Percentile	Confidence Level	Measured Ultimate Tensile Strength (PSI)	Estimated Lower Tolerance Bound on Yield Strength (PSI)
0.1%	95.0%	47,000	27,500
0.1%	95.0%	48,000	28,100
0.1%	95.0%	49,000	28,600
0.1%	95.0%	50,000	29,200
0.1%	95.0%	51,000	29,800
0.1%	95.0%	52,000	30,300
0.1%	95.0%	53,000	30,900
0.1%	95.0%	54,000	31,400
0.1%	95.0%	55,000	32,000
0.1%	95.0%	56,000	32,600
0.1%	95.0%	57,000	33,100
0.1%	95.0%	58,000	33,700
0.1%	95.0%	59,000	34,300
0.1%	95.0%	60,000	34,800
0.1%	95.0%	61,000	35,400
0.1%	95.0%	62,000	35,900
0.1%	95.0%	63,000	36,500
0.1%	95.0%	64,000	37,100
0.1%	95.0%	65,000	37,700
0.1%	95.0%	66,000	38,200
0.1%	95.0%	67,000	38,700
0.1%	95.0%	68,000	39,300
0.1%	95.0%	69,000	39,800
0.1%	95.0%	70,000	40,400
0.1%	95.0%	71,000	41,000
0.1%	95.0%	72,000	41,500
0.1%	95.0%	73,000	42,100
0.1%	95.0%	74,000	42,600
0.1%	95.0%	75,000	43,200
0.1%	95.0%	76,000	43,700
0.1%	95.0%	77,000	44,300
0.1%	95.0%	78,000	44,800
0.1%	95.0%	79,000	45,400
0.1%	95.0%	80,000	45,900
0.1%	95.0%	81,000	46,500
0.1%	95.0%	82,000	47,000
0.1%	95.0%	83,000	47,600
0.1%	95.0%	84,000	48,100
0.1%	95.0%	85,000	48,700
0.1%	95.0%	86,000	49,200
0.1%	95.0%	87,000	49,800
0.1%	95.0%	88,000	50,300
0.1%	95.0%	89,000	50,900
0.1%	95.0%	90,000	51,500
0.1%	95.0%	91,000	52,000
0.1%	95.0%	92,000	52,500
0.1%	95.0%	93,000	53,100
0.1%	95.0%	94,000	53,600
0.1%	95.0%	95,000	54,200
0.1%	95.0%	96,000	54,700
0.1%	95.0%	97,000	55,200
0.1%	95.0%	98,000	55,800
0.1%	95.0%	99,000	56,300
0.1%	95.0%	100,000	56,900

Table A.3.3. **Estimated lower tolerance bound yield strength as function of measured ultimate tensile strength, confidence level, and percentile — based on collected data with tensile property measures for transmission line joints manufactured prior to 1980 (Continued)**

Targeted Percentile	Confidence Level	Measured Ultimate Tensile Strength (PSI)	Estimated Lower Tolerance Bound on Yield Strength (PSI)
0.5%	99.0%	47,000	29,400
0.5%	99.0%	48,000	30,000
0.5%	99.0%	49,000	30,600
0.5%	99.0%	50,000	31,100
0.5%	99.0%	51,000	31,700
0.5%	99.0%	52,000	32,300
0.5%	99.0%	53,000	32,800
0.5%	99.0%	54,000	33,400
0.5%	99.0%	55,000	34,000
0.5%	99.0%	56,000	34,600
0.5%	99.0%	57,000	35,100
0.5%	99.0%	58,000	35,700
0.5%	99.0%	59,000	36,300
0.5%	99.0%	60,000	36,800
0.5%	99.0%	61,000	37,400
0.5%	99.0%	62,000	38,000
0.5%	99.0%	63,000	38,500
0.5%	99.0%	64,000	39,100
0.5%	99.0%	65,000	39,600
0.5%	99.0%	66,000	40,200
0.5%	99.0%	67,000	40,800
0.5%	99.0%	68,000	41,300
0.5%	99.0%	69,000	41,900
0.5%	99.0%	70,000	42,500
0.5%	99.0%	71,000	43,000
0.5%	99.0%	72,000	43,600
0.5%	99.0%	73,000	44,100
0.5%	99.0%	74,000	44,700
0.5%	99.0%	75,000	45,200
0.5%	99.0%	76,000	45,800
0.5%	99.0%	77,000	46,300
0.5%	99.0%	78,000	46,900
0.5%	99.0%	79,000	47,400
0.5%	99.0%	80,000	48,000
0.5%	99.0%	81,000	48,500
0.5%	99.0%	82,000	49,100
0.5%	99.0%	83,000	49,600
0.5%	99.0%	84,000	50,200
0.5%	99.0%	85,000	50,700
0.5%	99.0%	86,000	51,300
0.5%	99.0%	87,000	51,800
0.5%	99.0%	88,000	52,400
0.5%	99.0%	89,000	52,900
0.5%	99.0%	90,000	53,500
0.5%	99.0%	91,000	54,000
0.5%	99.0%	92,000	54,500
0.5%	99.0%	93,000	55,100
0.5%	99.0%	94,000	55,600
0.5%	99.0%	95,000	56,200
0.5%	99.0%	96,000	56,700
0.5%	99.0%	97,000	57,200
0.5%	99.0%	98,000	57,800
0.5%	99.0%	99,000	58,300
0.5%	99.0%	100,000	58,800

Table A.3.3. Estimated lower tolerance bound yield strength as function of measured ultimate tensile strength, confidence level, and percentile — based on collected data with tensile property measures for transmission line joints manufactured prior to 1980 (Continued)

Targeted Percentile	Confidence Level	Measured Ultimate Tensile Strength (PSI)	Estimated Lower Tolerance Bound on Yield Strength (PSI)
0.5%	95.0%	47,000	29,700
0.5%	95.0%	48,000	30,300
0.5%	95.0%	49,000	30,900
0.5%	95.0%	50,000	31,400
0.5%	95.0%	51,000	32,000
0.5%	95.0%	52,000	32,600
0.5%	95.0%	53,000	33,100
0.5%	95.0%	54,000	33,700
0.5%	95.0%	55,000	34,300
0.5%	95.0%	56,000	34,900
0.5%	95.0%	57,000	35,400
0.5%	95.0%	58,000	36,000
0.5%	95.0%	59,000	36,500
0.5%	95.0%	60,000	37,100
0.5%	95.0%	61,000	37,600
0.5%	95.0%	62,000	38,200
0.5%	95.0%	63,000	38,800
0.5%	95.0%	64,000	39,300
0.5%	95.0%	65,000	39,900
0.5%	95.0%	66,000	40,400
0.5%	95.0%	67,000	41,000
0.5%	95.0%	68,000	41,600
0.5%	95.0%	69,000	42,100
0.5%	95.0%	70,000	42,700
0.5%	95.0%	71,000	43,200
0.5%	95.0%	72,000	43,800
0.5%	95.0%	73,000	44,300
0.5%	95.0%	74,000	44,900
0.5%	95.0%	75,000	45,500
0.5%	95.0%	76,000	46,000
0.5%	95.0%	77,000	46,600
0.5%	95.0%	78,000	47,100
0.5%	95.0%	79,000	47,700
0.5%	95.0%	80,000	48,200
0.5%	95.0%	81,000	48,800
0.5%	95.0%	82,000	49,300
0.5%	95.0%	83,000	49,900
0.5%	95.0%	84,000	50,400
0.5%	95.0%	85,000	51,000
0.5%	95.0%	86,000	51,500
0.5%	95.0%	87,000	52,100
0.5%	95.0%	88,000	52,600
0.5%	95.0%	89,000	53,100
0.5%	95.0%	90,000	53,700
0.5%	95.0%	91,000	54,200
0.5%	95.0%	92,000	54,800
0.5%	95.0%	93,000	55,300
0.5%	95.0%	94,000	55,900
0.5%	95.0%	95,000	56,400
0.5%	95.0%	96,000	57,000
0.5%	95.0%	97,000	57,500
0.5%	95.0%	98,000	58,100
0.5%	95.0%	99,000	58,600
0.5%	95.0%	100,000	59,100

Table A.3.3. Estimated lower tolerance bound yield strength as function of measured ultimate tensile strength, confidence level, and percentile — based on collected data with tensile property measures for transmission line joints manufactured prior to 1980 (Continued)

Targeted Percentile	Confidence Level	Measured Ultimate Tensile Strength (PSI)	Estimated Lower Tolerance Bound on Yield Strength (PSI)
1.0%	99.9%	47,000	30,200
1.0%	99.9%	48,000	30,800
1.0%	99.9%	49,000	31,300
1.0%	99.9%	50,000	31,900
1.0%	99.9%	51,000	32,500
1.0%	99.9%	52,000	33,100
1.0%	99.9%	53,000	33,600
1.0%	99.9%	54,000	34,200
1.0%	99.9%	55,000	34,800
1.0%	99.9%	56,000	35,400
1.0%	99.9%	57,000	35,900
1.0%	99.9%	58,000	36,500
1.0%	99.9%	59,000	37,100
1.0%	99.9%	60,000	37,600
1.0%	99.9%	61,000	38,200
1.0%	99.9%	62,000	38,800
1.0%	99.9%	63,000	39,400
1.0%	99.9%	64,000	39,900
1.0%	99.9%	65,000	40,500
1.0%	99.9%	66,000	41,100
1.0%	99.9%	67,000	41,600
1.0%	99.9%	68,000	42,200
1.0%	99.9%	69,000	42,800
1.0%	99.9%	70,000	43,300
1.0%	99.9%	71,000	43,900
1.0%	99.9%	72,000	44,500
1.0%	99.9%	73,000	45,000
1.0%	99.9%	74,000	45,600
1.0%	99.9%	75,000	46,100
1.0%	99.9%	76,000	46,700
1.0%	99.9%	77,000	47,200
1.0%	99.9%	78,000	47,800
1.0%	99.9%	79,000	48,300
1.0%	99.9%	80,000	48,900
1.0%	99.9%	81,000	49,400
1.0%	99.9%	82,000	50,000
1.0%	99.9%	83,000	50,500
1.0%	99.9%	84,000	51,100
1.0%	99.9%	85,000	51,600
1.0%	99.9%	86,000	52,200
1.0%	99.9%	87,000	52,700
1.0%	99.9%	88,000	53,200
1.0%	99.9%	89,000	53,800
1.0%	99.9%	90,000	54,300
1.0%	99.9%	91,000	54,900
1.0%	99.9%	92,000	55,400
1.0%	99.9%	93,000	55,900
1.0%	99.9%	94,000	56,500
1.0%	99.9%	95,000	57,000
1.0%	99.9%	96,000	57,500
1.0%	99.9%	97,000	58,100
1.0%	99.9%	98,000	58,600
1.0%	99.9%	99,000	59,100
1.0%	99.9%	100,000	59,600

Table A.3.3. Estimated lower tolerance bound yield strength as function of measured ultimate tensile strength, confidence level, and percentile — based on collected data with tensile property measures for transmission line joints manufactured prior to 1980 (Continued)

Targeted Percentile	Confidence Level	Measured Ultimate Tensile Strength (PSI)	Estimated Lower Tolerance Bound on Yield Strength (PSI)
1.0%	99.0%	47,000	30,500
1.0%	99.0%	48,000	31,100
1.0%	99.0%	49,000	31,700
1.0%	99.0%	50,000	32,200
1.0%	99.0%	51,000	32,800
1.0%	99.0%	52,000	33,400
1.0%	99.0%	53,000	34,000
1.0%	99.0%	54,000	34,500
1.0%	99.0%	55,000	35,100
1.0%	99.0%	56,000	35,700
1.0%	99.0%	57,000	36,200
1.0%	99.0%	58,000	36,800
1.0%	99.0%	59,000	37,400
1.0%	99.0%	60,000	37,900
1.0%	99.0%	61,000	38,500
1.0%	99.0%	62,000	39,100
1.0%	99.0%	63,000	39,600
1.0%	99.0%	64,000	40,200
1.0%	99.0%	65,000	40,800
1.0%	99.0%	66,000	41,300
1.0%	99.0%	67,000	41,900
1.0%	99.0%	68,000	42,400
1.0%	99.0%	69,000	43,000
1.0%	99.0%	70,000	43,500
1.0%	99.0%	71,000	44,100
1.0%	99.0%	72,000	44,700
1.0%	99.0%	73,000	45,200
1.0%	99.0%	74,000	45,800
1.0%	99.0%	75,000	46,400
1.0%	99.0%	76,000	46,900
1.0%	99.0%	77,000	47,500
1.0%	99.0%	78,000	48,000
1.0%	99.0%	79,000	48,600
1.0%	99.0%	80,000	49,100
1.0%	99.0%	81,000	49,700
1.0%	99.0%	82,000	50,200
1.0%	99.0%	83,000	50,800
1.0%	99.0%	84,000	51,400
1.0%	99.0%	85,000	51,900
1.0%	99.0%	86,000	52,400
1.0%	99.0%	87,000	52,900
1.0%	99.0%	88,000	53,500
1.0%	99.0%	89,000	54,000
1.0%	99.0%	90,000	54,600
1.0%	99.0%	91,000	55,100
1.0%	99.0%	92,000	55,700
1.0%	99.0%	93,000	56,200
1.0%	99.0%	94,000	56,700
1.0%	99.0%	95,000	57,200
1.0%	99.0%	96,000	57,800
1.0%	99.0%	97,000	58,300
1.0%	99.0%	98,000	58,900
1.0%	99.0%	99,000	59,400
1.0%	99.0%	100,000	60,000

Table A.3.3. Estimated lower tolerance bound yield strength as function of measured ultimate tensile strength, confidence level, and percentile — based on collected data with tensile property measures for transmission line joints manufactured prior to 1980 (Continued)

Targeted Percentile	Confidence Level	Measured Ultimate Tensile Strength (PSI)	Estimated Lower Tolerance Bound on Yield Strength (PSI)
1.0%	95.0%	47,000	30,800
1.0%	95.0%	48,000	31,400
1.0%	95.0%	49,000	32,000
1.0%	95.0%	50,000	32,500
1.0%	95.0%	51,000	33,100
1.0%	95.0%	52,000	33,700
1.0%	95.0%	53,000	34,200
1.0%	95.0%	54,000	34,800
1.0%	95.0%	55,000	35,400
1.0%	95.0%	56,000	35,900
1.0%	95.0%	57,000	36,500
1.0%	95.0%	58,000	37,000
1.0%	95.0%	59,000	37,600
1.0%	95.0%	60,000	38,200
1.0%	95.0%	61,000	38,700
1.0%	95.0%	62,000	39,300
1.0%	95.0%	63,000	39,900
1.0%	95.0%	64,000	40,400
1.0%	95.0%	65,000	41,000
1.0%	95.0%	66,000	41,500
1.0%	95.0%	67,000	42,100
1.0%	95.0%	68,000	42,700
1.0%	95.0%	69,000	43,300
1.0%	95.0%	70,000	43,800
1.0%	95.0%	71,000	44,300
1.0%	95.0%	72,000	44,900
1.0%	95.0%	73,000	45,400
1.0%	95.0%	74,000	46,000
1.0%	95.0%	75,000	46,600
1.0%	95.0%	76,000	47,100
1.0%	95.0%	77,000	47,700
1.0%	95.0%	78,000	48,200
1.0%	95.0%	79,000	48,800
1.0%	95.0%	80,000	49,300
1.0%	95.0%	81,000	49,900
1.0%	95.0%	82,000	50,400
1.0%	95.0%	83,000	51,000
1.0%	95.0%	84,000	51,500
1.0%	95.0%	85,000	52,100
1.0%	95.0%	86,000	52,600
1.0%	95.0%	87,000	53,200
1.0%	95.0%	88,000	53,700
1.0%	95.0%	89,000	54,200
1.0%	95.0%	90,000	54,800
1.0%	95.0%	91,000	55,300
1.0%	95.0%	92,000	55,900
1.0%	95.0%	93,000	56,400
1.0%	95.0%	94,000	57,000
1.0%	95.0%	95,000	57,500
1.0%	95.0%	96,000	58,100
1.0%	95.0%	97,000	58,600
1.0%	95.0%	98,000	59,100
1.0%	95.0%	99,000	59,700
1.0%	95.0%	100,000	60,200

Table A.3.6. Estimated lower tolerance bound ultimate tensile strength as a function of measured Rockwell B hardness, confidence level, and percentile — based on collected data with tensile and hardness property measures for transmission line joints manufactured prior to 1980

Targeted Percentile	Confidence Level	Measured Hardness (HRB)	Estimated Lower Tolerance Bound on Tensile Strength (PSI)
0.1%	99.9%	54	25,400
0.1%	99.9%	55	26,300
0.1%	99.9%	56	27,200
0.1%	99.9%	57	28,100
0.1%	99.9%	58	29,000
0.1%	99.9%	59	29,900
0.1%	99.9%	60	30,700
0.1%	99.9%	61	31,600
0.1%	99.9%	62	32,500
0.1%	99.9%	63	33,400
0.1%	99.9%	64	34,300
0.1%	99.9%	65	35,200
0.1%	99.9%	66	36,100
0.1%	99.9%	67	36,900
0.1%	99.9%	68	37,800
0.1%	99.9%	69	38,700
0.1%	99.9%	70	39,600
0.1%	99.9%	71	40,400
0.1%	99.9%	72	41,300
0.1%	99.9%	73	42,200
0.1%	99.9%	74	43,000
0.1%	99.9%	75	43,900
0.1%	99.9%	76	44,700
0.1%	99.9%	77	45,500
0.1%	99.9%	78	46,400
0.1%	99.9%	79	47,200
0.1%	99.9%	80	48,000
0.1%	99.9%	81	48,800
0.1%	99.9%	82	49,600
0.1%	99.9%	83	50,400
0.1%	99.9%	84	51,100
0.1%	99.9%	85	51,900
0.1%	99.9%	86	52,600
0.1%	99.9%	87	53,300
0.1%	99.9%	88	54,000
0.1%	99.9%	89	54,700
0.1%	99.9%	90	55,400
0.1%	99.9%	91	56,100
0.1%	99.9%	92	56,700
0.1%	99.9%	93	57,400
0.1%	99.9%	94	58,000
0.1%	99.9%	95	58,600
0.1%	99.9%	96	59,300
0.1%	99.9%	97	59,900

Table A.3.6. Estimated lower tolerance bound ultimate tensile strength as function of measured Rockwell B hardness, confidence level, and percentile — based on collected data with tensile and hardness property measures for transmission line joints manufactured prior to 1980 (Continued)

Targeted Percentile	Confidence Level	Measured Hardness (HRB)	Estimated Lower Tolerance Bound on Tensile Strength (PSI)
0.1%	99.0%	54	26,900
0.1%	99.0%	55	27,800
0.1%	99.0%	56	28,700
0.1%	99.0%	57	29,500
0.1%	99.0%	58	30,400
0.1%	99.0%	59	31,200
0.1%	99.0%	60	32,100
0.1%	99.0%	61	32,900
0.1%	99.0%	62	33,800
0.1%	99.0%	63	34,600
0.1%	99.0%	64	35,500
0.1%	99.0%	65	36,300
0.1%	99.0%	66	37,100
0.1%	99.0%	67	38,000
0.1%	99.0%	68	38,800
0.1%	99.0%	69	39,700
0.1%	99.0%	70	40,500
0.1%	99.0%	71	41,300
0.1%	99.0%	72	42,200
0.1%	99.0%	73	43,000
0.1%	99.0%	74	43,800
0.1%	99.0%	75	44,600
0.1%	99.0%	76	45,500
0.1%	99.0%	77	46,300
0.1%	99.0%	78	47,100
0.1%	99.0%	79	47,900
0.1%	99.0%	80	48,700
0.1%	99.0%	81	49,400
0.1%	99.0%	82	50,200
0.1%	99.0%	83	51,000
0.1%	99.0%	84	51,700
0.1%	99.0%	85	52,500
0.1%	99.0%	86	53,200
0.1%	99.0%	87	53,900
0.1%	99.0%	88	54,700
0.1%	99.0%	89	55,400
0.1%	99.0%	90	56,000
0.1%	99.0%	91	56,700
0.1%	99.0%	92	57,400
0.1%	99.0%	93	58,100
0.1%	99.0%	94	58,700
0.1%	99.0%	95	59,400
0.1%	99.0%	96	60,000
0.1%	99.0%	97	60,700

Table A.3.6. Estimated lower tolerance bound ultimate tensile strength as function of measured Rockwell B hardness, confidence level, and percentile — based on collected data with tensile and hardness property measures for transmission line joints manufactured prior to 1980 (Continued)

Targeted Percentile	Confidence Level	Measured Hardness (HRB)	Estimated Lower Tolerance Bound on Tensile Strength (PSI)
0.1%	95.0%	54	28,300
0.1%	95.0%	55	29,100
0.1%	95.0%	56	29,900
0.1%	95.0%	57	30,800
0.1%	95.0%	58	31,600
0.1%	95.0%	59	32,400
0.1%	95.0%	60	33,200
0.1%	95.0%	61	34,000
0.1%	95.0%	62	34,800
0.1%	95.0%	63	35,700
0.1%	95.0%	64	36,500
0.1%	95.0%	65	37,300
0.1%	95.0%	66	38,100
0.1%	95.0%	67	38,900
0.1%	95.0%	68	39,700
0.1%	95.0%	69	40,500
0.1%	95.0%	70	41,300
0.1%	95.0%	71	42,100
0.1%	95.0%	72	42,900
0.1%	95.0%	73	43,700
0.1%	95.0%	74	44,500
0.1%	95.0%	75	45,300
0.1%	95.0%	76	46,100
0.1%	95.0%	77	46,900
0.1%	95.0%	78	47,700
0.1%	95.0%	79	48,500
0.1%	95.0%	80	49,200
0.1%	95.0%	81	50,000
0.1%	95.0%	82	50,800
0.1%	95.0%	83	51,600
0.1%	95.0%	84	52,300
0.1%	95.0%	85	53,000
0.1%	95.0%	86	53,700
0.1%	95.0%	87	54,500
0.1%	95.0%	88	55,200
0.1%	95.0%	89	55,900
0.1%	95.0%	90	56,600
0.1%	95.0%	91	57,300
0.1%	95.0%	92	58,000
0.1%	95.0%	93	58,700
0.1%	95.0%	94	59,400
0.1%	95.0%	95	60,000
0.1%	95.0%	96	60,700
0.1%	95.0%	97	61,400

Table A.3.6. Estimated lower tolerance bound ultimate tensile strength as function of measured Rockwell B hardness, confidence level, and percentile — based on collected data with tensile and hardness property measures for transmission line joints manufactured prior to 1980 (Continued)

Targeted Percentile	Confidence Level	Measured Hardness (HRB)	Estimated Lower Tolerance Bound on Tensile Strength (PSI)
0.5%	99.9%	54	29,000
0.5%	99.9%	55	29,900
0.5%	99.9%	56	30,800
0.5%	99.9%	57	31,700
0.5%	99.9%	58	32,600
0.5%	99.9%	59	33,500
0.5%	99.9%	60	34,400
0.5%	99.9%	61	35,300
0.5%	99.9%	62	36,200
0.5%	99.9%	63	37,100
0.5%	99.9%	64	37,900
0.5%	99.9%	65	38,800
0.5%	99.9%	66	39,700
0.5%	99.9%	67	40,600
0.5%	99.9%	68	41,500
0.5%	99.9%	69	42,400
0.5%	99.9%	70	43,200
0.5%	99.9%	71	44,100
0.5%	99.9%	72	45,000
0.5%	99.9%	73	45,900
0.5%	99.9%	74	46,700
0.5%	99.9%	75	47,600
0.5%	99.9%	76	48,400
0.5%	99.9%	77	49,300
0.5%	99.9%	78	50,100
0.5%	99.9%	79	51,000
0.5%	99.9%	80	51,800
0.5%	99.9%	81	52,600
0.5%	99.9%	82	53,400
0.5%	99.9%	83	54,200
0.5%	99.9%	84	54,900
0.5%	99.9%	85	55,700
0.5%	99.9%	86	56,400
0.5%	99.9%	87	57,100
0.5%	99.9%	88	57,800
0.5%	99.9%	89	58,500
0.5%	99.9%	90	59,200
0.5%	99.9%	91	59,900
0.5%	99.9%	92	60,500
0.5%	99.9%	93	61,100
0.5%	99.9%	94	61,800
0.5%	99.9%	95	62,400
0.5%	99.9%	96	63,000
0.5%	99.9%	97	63,700

Table A.3.6. Estimated lower tolerance bound ultimate tensile strength as function of measured Rockwell B hardness, confidence level, and percentile — based on collected data with tensile and hardness property measures for transmission line joints manufactured prior to 1980 (Continued)

Targeted Percentile	Confidence Level	Measured Hardness (HRB)	Estimated Lower Tolerance Bound on Tensile Strength (PSI)
0.5%	99.0%	54	30,500
0.5%	99.0%	55	31,400
0.5%	99.0%	56	32,200
0.5%	99.0%	57	33,100
0.5%	99.0%	58	33,900
0.5%	99.0%	59	34,800
0.5%	99.0%	60	35,700
0.5%	99.0%	61	36,500
0.5%	99.0%	62	37,400
0.5%	99.0%	63	38,200
0.5%	99.0%	64	39,100
0.5%	99.0%	65	39,900
0.5%	99.0%	66	40,800
0.5%	99.0%	67	41,600
0.5%	99.0%	68	42,400
0.5%	99.0%	69	43,300
0.5%	99.0%	70	44,100
0.5%	99.0%	71	45,000
0.5%	99.0%	72	45,800
0.5%	99.0%	73	46,600
0.5%	99.0%	74	47,400
0.5%	99.0%	75	48,300
0.5%	99.0%	76	49,100
0.5%	99.0%	77	49,900
0.5%	99.0%	78	50,800
0.5%	99.0%	79	51,600
0.5%	99.0%	80	52,400
0.5%	99.0%	81	53,100
0.5%	99.0%	82	53,900
0.5%	99.0%	83	54,700
0.5%	99.0%	84	55,500
0.5%	99.0%	85	56,200
0.5%	99.0%	86	56,900
0.5%	99.0%	87	57,700
0.5%	99.0%	88	58,400
0.5%	99.0%	89	59,100
0.5%	99.0%	90	59,700
0.5%	99.0%	91	60,400
0.5%	99.0%	92	61,100
0.5%	99.0%	93	61,800
0.5%	99.0%	94	62,400
0.5%	99.0%	95	63,100
0.5%	99.0%	96	63,700
0.5%	99.0%	97	64,400

Table A.3.6. Estimated lower tolerance bound ultimate tensile strength as function of measured Rockwell B hardness, confidence level, and percentile — based on collected data with tensile and hardness property measures for transmission line joints manufactured prior to 1980 (Continued)

Targeted Percentile	Confidence Level	Measured Hardness (HRB)	Estimated Lower Tolerance Bound on Tensile Strength (PSI)
0.5%	95.0%	54	31,900
0.5%	95.0%	55	32,700
0.5%	95.0%	56	33,500
0.5%	95.0%	57	34,300
0.5%	95.0%	58	35,100
0.5%	95.0%	59	36,000
0.5%	95.0%	60	36,800
0.5%	95.0%	61	37,600
0.5%	95.0%	62	38,400
0.5%	95.0%	63	39,200
0.5%	95.0%	64	40,000
0.5%	95.0%	65	40,900
0.5%	95.0%	66	41,700
0.5%	95.0%	67	42,500
0.5%	95.0%	68	43,300
0.5%	95.0%	69	44,100
0.5%	95.0%	70	44,900
0.5%	95.0%	71	45,700
0.5%	95.0%	72	46,500
0.5%	95.0%	73	47,300
0.5%	95.0%	74	48,100
0.5%	95.0%	75	48,900
0.5%	95.0%	76	49,700
0.5%	95.0%	77	50,500
0.5%	95.0%	78	51,300
0.5%	95.0%	79	52,100
0.5%	95.0%	80	52,900
0.5%	95.0%	81	53,600
0.5%	95.0%	82	54,400
0.5%	95.0%	83	55,200
0.5%	95.0%	84	55,900
0.5%	95.0%	85	56,700
0.5%	95.0%	86	57,400
0.5%	95.0%	87	58,100
0.5%	95.0%	88	58,800
0.5%	95.0%	89	59,500
0.5%	95.0%	90	60,200
0.5%	95.0%	91	60,900
0.5%	95.0%	92	61,600
0.5%	95.0%	93	62,300
0.5%	95.0%	94	63,000
0.5%	95.0%	95	63,600
0.5%	95.0%	96	64,300
0.5%	95.0%	97	65,000

Table A.3.6. Estimated lower tolerance bound ultimate tensile strength as function of measured Rockwell B hardness, confidence level, and percentile — based on collected data with tensile and hardness property measures for transmission line joints manufactured prior to 1980 (Continued)

Targeted Percentile	Confidence Level	Measured Hardness (HRB)	Estimated Lower Tolerance Bound on Tensile Strength (PSI)
1.0%	99.9%	54	30,700
1.0%	99.9%	55	31,600
1.0%	99.9%	56	32,500
1.0%	99.9%	57	33,400
1.0%	99.9%	58	34,300
1.0%	99.9%	59	35,200
1.0%	99.9%	60	36,100
1.0%	99.9%	61	37,000
1.0%	99.9%	62	37,900
1.0%	99.9%	63	38,800
1.0%	99.9%	64	39,700
1.0%	99.9%	65	40,600
1.0%	99.9%	66	41,500
1.0%	99.9%	67	42,400
1.0%	99.9%	68	43,300
1.0%	99.9%	69	44,100
1.0%	99.9%	70	45,000
1.0%	99.9%	71	45,900
1.0%	99.9%	72	46,800
1.0%	99.9%	73	47,700
1.0%	99.9%	74	48,500
1.0%	99.9%	75	49,400
1.0%	99.9%	76	50,300
1.0%	99.9%	77	51,100
1.0%	99.9%	78	51,900
1.0%	99.9%	79	52,800
1.0%	99.9%	80	53,600
1.0%	99.9%	81	54,400
1.0%	99.9%	82	55,200
1.0%	99.9%	83	56,000
1.0%	99.9%	84	56,800
1.0%	99.9%	85	57,500
1.0%	99.9%	86	58,300
1.0%	99.9%	87	59,000
1.0%	99.9%	88	59,700
1.0%	99.9%	89	60,300
1.0%	99.9%	90	61,000
1.0%	99.9%	91	61,700
1.0%	99.9%	92	62,300
1.0%	99.9%	93	62,900
1.0%	99.9%	94	63,600
1.0%	99.9%	95	64,200
1.0%	99.9%	96	64,800
1.0%	99.9%	97	65,400

Table A.3.6. Estimated lower tolerance bound ultimate tensile strength as function of measured Rockwell B hardness, confidence level, and percentile — based on collected data with tensile and hardness property measures for transmission line joints manufactured prior to 1980 (Continued)

Targeted Percentile	Confidence Level	Measured Hardness (HRB)	Estimated Lower Tolerance Bound on Tensile Strength (PSI)
1.0%	99.0%	54	32,300
1.0%	99.0%	55	33,100
1.0%	99.0%	56	34,000
1.0%	99.0%	57	34,800
1.0%	99.0%	58	35,700
1.0%	99.0%	59	36,500
1.0%	99.0%	60	37,400
1.0%	99.0%	61	38,200
1.0%	99.0%	62	39,100
1.0%	99.0%	63	39,900
1.0%	99.0%	64	40,800
1.0%	99.0%	65	41,700
1.0%	99.0%	66	42,500
1.0%	99.0%	67	43,400
1.0%	99.0%	68	44,200
1.0%	99.0%	69	45,000
1.0%	99.0%	70	45,900
1.0%	99.0%	71	46,700
1.0%	99.0%	72	47,600
1.0%	99.0%	73	48,400
1.0%	99.0%	74	49,200
1.0%	99.0%	75	50,100
1.0%	99.0%	76	50,900
1.0%	99.0%	77	51,700
1.0%	99.0%	78	52,500
1.0%	99.0%	79	53,300
1.0%	99.0%	80	54,100
1.0%	99.0%	81	54,900
1.0%	99.0%	82	55,700
1.0%	99.0%	83	56,500
1.0%	99.0%	84	57,300
1.0%	99.0%	85	58,000
1.0%	99.0%	86	58,700
1.0%	99.0%	87	59,500
1.0%	99.0%	88	60,200
1.0%	99.0%	89	60,900
1.0%	99.0%	90	61,600
1.0%	99.0%	91	62,200
1.0%	99.0%	92	62,900
1.0%	99.0%	93	63,500
1.0%	99.0%	94	64,200
1.0%	99.0%	95	64,800
1.0%	99.0%	96	65,500
1.0%	99.0%	97	66,100

Table A.3.6. Estimated lower tolerance bound ultimate tensile strength as function of measured Rockwell B hardness, confidence level, and percentile — based on collected data with tensile and hardness property measures for transmission line joints manufactured prior to 1980 (Continued)

Targeted Percentile	Confidence Level	Measured Hardness (HRB)	Estimated Lower Tolerance Bound on Tensile Strength (PSI)
1.0%	95.0%	54	33,600
1.0%	95.0%	55	34,400.
1.0%	95.0%	56	35,200
1.0%	95.0%	57	36,000
1.0%	95.0%	58	36,900
1.0%	95.0%	59	37,700
1.0%	95.0%	60	38,500
1.0%	95.0%	61	39,300
1.0%	95.0%	62	40,100
1.0%	95.0%	63	40,900
1.0%	95.0%	64	41,800
1.0%	95.0%	65	42,600
1.0%	95.0%	66	43,400
1.0%	95.0%	67	44,200
1.0%	95.0%	68	45,000
1.0%	95.0%	69	45,800
1.0%	95.0%	70	46,600
1.0%	95.0%	71	47,500
1.0%	95.0%	72	48,300
1.0%	95.0%	73	49,100
1.0%	95.0%	74	49,900
1.0%	95.0%	75	50,700
1.0%	95.0%	76	51,500
1.0%	95.0%	77	52,300
1.0%	95.0%	78	53,100
1.0%	95.0%	79	53,800
1.0%	95.0%	80	54,600
1.0%	95.0%	81	55,400
1.0%	95.0%	82	56,200
1.0%	95.0%	83	56,900
1.0%	95.0%	84	57,700
1.0%	95.0%	85	58,400
1.0%	95.0%	86	59,200
1.0%	95.0%	87	59,900
1.0%	95.0%	88	60,600
1.0%	95.0%	89	61,300
1.0%	95.0%	90	62,000
1.0%	95.0%	91	62,700
1.0%	95.0%	92	63,400
1.0%	95.0%	93	64,000
1.0%	95.0%	94	64,700
1.0%	95.0%	95	65,400
1.0%	95.0%	96	66,100
1.0%	95.0%	97	66,700

Table A.3.8. Estimated lower tolerance bound yield strength as a function of Rockwell B hardness, confidence level, and percentile — based on collected data with tensile and hardness property measures for transmission line joints manufactured prior to 1980

Targeted Percentile	Confidence Level	Measured Hardness (HRB)	Estimated Lower Tolerance Bound on Yield Strength (PSI)
0.1%	99.9%	54	17,900
0.1%	99.9%	55	18,700
0.1%	99.9%	56	19,500
0.1%	99.9%	57	20,300
0.1%	99.9%	58	21,100
0.1%	99.9%	59	21,900
0.1%	99.9%	60	22,700
0.1%	99.9%	61	23,500
0.1%	99.9%	62	24,300
0.1%	99.9%	63	25,100
0.1%	99.9%	64	25,900
0.1%	99.9%	65	26,700
0.1%	99.9%	66	27,500
0.1%	99.9%	67	28,300
0.1%	99.9%	68	29,100
0.1%	99.9%	69	29,900
0.1%	99.9%	70	30,700
0.1%	99.9%	71	31,500
0.1%	99.9%	72	32,300
0.1%	99.9%	73	33,000
0.1%	99.9%	74	33,800
0.1%	99.9%	75	34,600
0.1%	99.9%	76	35,400
0.1%	99.9%	77	36,100
0.1%	99.9%	78	36,900
0.1%	99.9%	79	37,700
0.1%	99.9%	80	38,400
0.1%	99.9%	81	39,200
0.1%	99.9%	82	39,900
0.1%	99.9%	83	40,600
0.1%	99.9%	84	41,400
0.1%	99.9%	85	42,100
0.1%	99.9%	86	42,800
0.1%	99.9%	87	43,500
0.1%	99.9%	88	44,100
0.1%	99.9%	89	44,800
0.1%	99.9%	90	45,500
0.1%	99.9%	91	46,100
0.1%	99.9%	92	46,800
0.1%	99.9%	93	47,400
0.1%	99.9%	94	48,000
0.1%	99.9%	95	48,700
0.1%	99.9%	96	49,300
0.1%	99.9%	97	49,900

Table A.3.8. Estimated lower tolerance bound yield strength as function of Rockwell B hardness, confidence level, and percentile — based on collected data with tensile and hardness property measures for transmission line joints manufactured prior to 1980 (Continued).

Targeted Percentile	Confidence Level	Measured Hardness (HRB)	Estimated Lower Tolerance Bound on Yield Strength (PSI)
0.1%	99.0%	54	18,900
0.1%	99.0%	55	19,700
0.1%	99.0%	56	20,500
0.1%	99.0%	57	21,300
0.1%	99.0%	58	22,000
0.1%	99.0%	59	22,800
0.1%	99.0%	60	23,600
0.1%	99.0%	61	24,400
0.1%	99.0%	62	25,100
0.1%	99.0%	63	25,900
0.1%	99.0%	64	26,700
0.1%	99.0%	65	27,500
0.1%	99.0%	66	28,200
0.1%	99.0%	67	29,000
0.1%	99.0%	68	29,800
0.1%	99.0%	69	30,600
0.1%	99.0%	70	31,300
0.1%	99.0%	71	32,100
0.1%	99.0%	72	32,900
0.1%	99.0%	73	33,600
0.1%	99.0%	74	34,400
0.1%	99.0%	75	35,100
0.1%	99.0%	76	35,900
0.1%	99.0%	77	36,600
0.1%	99.0%	78	37,400
0.1%	99.0%	79	38,100
0.1%	99.0%	80	38,900
0.1%	99.0%	81	39,600
0.1%	99.0%	82	40,300
0.1%	99.0%	83	41,100
0.1%	99.0%	84	41,800
0.1%	99.0%	85	42,500
0.1%	99.0%	86	43,200
0.1%	99.0%	87	43,800
0.1%	99.0%	88	44,500
0.1%	99.0%	89	45,200
0.1%	99.0%	90	45,900
0.1%	99.0%	91	46,500
0.1%	99.0%	92	47,200
0.1%	99.0%	93	47,900
0.1%	99.0%	94	48,500
0.1%	99.0%	95	49,200
0.1%	99.0%	96	49,800
0.1%	99.0%	97	50,400

Table A.3.8. Estimated lower tolerance bound yield strength as function of Rockwell B hardness, confidence level, and percentile — based on collected data with tensile and hardness property measures for transmission line joints manufactured prior to 1980 (Continued).

Targeted Percentile	Confidence Level	Measured Hardness (HRB)	Estimated Lower Tolerance Bound on Yield Strength (PSI)
0.1%	95.0%	54	19,800
0.1%	95.0%	55	20,600
0.1%	95.0%	56	21,300
0.1%	95.0%	57	22,100
0.1%	95.0%	58	22,900
0.1%	95.0%	59	23,600
0.1%	95.0%	60	24,400
0.1%	95.0%	61	25,100
0.1%	95.0%	62	25,900
0.1%	95.0%	63	26,600
0.1%	95.0%	64	27,400
0.1%	95.0%	65	28,100
0.1%	95.0%	66	28,900
0.1%	95.0%	67	29,600
0.1%	95.0%	68	30,400
0.1%	95.0%	69	31,100
0.1%	95.0%	70	31,900
0.1%	95.0%	71	32,600
0.1%	95.0%	72	33,400
0.1%	95.0%	73	34,100
0.1%	95.0%	74	34,800
0.1%	95.0%	75	35,600
0.1%	95.0%	76	36,300
0.1%	95.0%	77	37,100
0.1%	95.0%	78	37,800
0.1%	95.0%	79	38,500
0.1%	95.0%	80	39,200
0.1%	95.0%	81	40,000
0.1%	95.0%	82	40,700
0.1%	95.0%	83	41,400
0.1%	95.0%	84	42,100
0.1%	95.0%	85	42,800
0.1%	95.0%	86	43,500
0.1%	95.0%	87	44,200
0.1%	95.0%	88	44,900
0.1%	95.0%	89	45,600
0.1%	95.0%	90	46,300
0.1%	95.0%	91	46,900
0.1%	95.0%	92	47,600
0.1%	95.0%	93	48,300
0.1%	95.0%	94	48,900
0.1%	95.0%	95	49,600
0.1%	95.0%	96	50,200
0.1%	95.0%	97	50,900

Table A.3.8. Estimated lower tolerance bound yield strength as function of Rockwell B hardness, confidence level, and percentile — based on collected data with tensile and hardness property measures for transmission line joints manufactured prior to 1980 (Continued).

Targeted Percentile	Confidence Level	Measured Hardness (HRB)	Estimated Lower Tolerance Bound on Yield Strength (PSI)
0.5%	99.9%	54	20,300
0.5%	99.9%	55	21,100
0.5%	99.9%	56	21,900
0.5%	99.9%	57	22,700
0.5%	99.9%	58	23,500
0.5%	99.9%	59	24,300
0.5%	99.9%	60	25,100
0.5%	99.9%	61	25,900
0.5%	99.9%	62	26,800
0.5%	99.9%	63	27,600
0.5%	99.9%	64	28,400
0.5%	99.9%	65	29,200
0.5%	99.9%	66	30,000
0.5%	99.9%	67	30,800
0.5%	99.9%	68	31,600
0.5%	99.9%	69	32,400
0.5%	99.9%	70	33,200
0.5%	99.9%	71	34,000
0.5%	99.9%	72	34,700
0.5%	99.9%	73	35,500
0.5%	99.9%	74	36,300
0.5%	99.9%	75	37,100
0.5%	99.9%	76	37,900
0.5%	99.9%	77	38,700
0.5%	99.9%	78	39,400
0.5%	99.9%	79	40,200
0.5%	99.9%	80	40,900
0.5%	99.9%	81	41,700
0.5%	99.9%	82	42,400
0.5%	99.9%	83	43,200
0.5%	99.9%	84	43,900
0.5%	99.9%	85	44,600
0.5%	99.9%	86	45,300
0.5%	99.9%	87	46,000
0.5%	99.9%	88	46,700
0.5%	99.9%	89	47,300
0.5%	99.9%	90	48,000
0.5%	99.9%	91	48,700
0.5%	99.9%	92	49,300
0.5%	99.9%	93	49,900
0.5%	99.9%	94	50,500
0.5%	99.9%	95	51,200
0.5%	99.9%	96	51,800
0.5%	99.9%	97	52,400

Table A.3.8. Estimated lower tolerance bound yield strength as function of Rockwell B hardness, confidence level, and percentile — based on collected data with tensile and hardness property measures for transmission line joints manufactured prior to 1980 (Continued).

Targeted Percentile	Confidence Level	Measured Hardness (HRB)	Estimated Lower Tolerance Bound on Yield Strength (PSI)
0.5%	99.0%	54	21,300
0.5%	99.0%	55	22,100
0.5%	99.0%	56	22,900
0.5%	99.0%	57	23,700
0.5%	99.0%	58	24,400
0.5%	99.0%	59	25,200
0.5%	99.0%	60	26,000
0.5%	99.0%	61	26,800
0.5%	99.0%	62	27,600
0.5%	99.0%	63	28,300
0.5%	99.0%	64	29,100
0.5%	99.0%	65	29,900
0.5%	99.0%	66	30,700
0.5%	99.0%	67	31,400
0.5%	99.0%	68	32,200
0.5%	99.0%	69	33,000
0.5%	99.0%	70	33,800
0.5%	99.0%	71	34,500
0.5%	99.0%	72	35,300
0.5%	99.0%	73	36,100
0.5%	99.0%	74	36,800
0.5%	99.0%	75	37,600
0.5%	99.0%	76	38,300
0.5%	99.0%	77	39,100
0.5%	99.0%	78	39,900
0.5%	99.0%	79	40,600
0.5%	99.0%	80	41,400
0.5%	99.0%	81	42,100
0.5%	99.0%	82	42,800
0.5%	99.0%	83	43,500
0.5%	99.0%	84	44,200
0.5%	99.0%	85	45,000
0.5%	99.0%	86	45,700
0.5%	99.0%	87	46,300
0.5%	99.0%	88	47,000
0.5%	99.0%	89	47,700
0.5%	99.0%	90	48,400
0.5%	99.0%	91	49,100
0.5%	99.0%	92	49,700
0.5%	99.0%	93	50,300
0.5%	99.0%	94	51,000
0.5%	99.0%	95	51,600
0.5%	99.0%	96	52,300
0.5%	99.0%	97	52,900

Table A.3.8. Estimated lower tolerance bound yield strength as function of Rockwell B hardness, confidence level, and percentile — based on collected data with tensile and hardness property measures for transmission line joints manufactured prior to 1980 (Continued).

Targeted Percentile	Confidence Level	Measured Hardness (HRB)	Estimated Lower Tolerance Bound on Yield Strength (PSI)
0.5%	95.0%	54	$22,200
0.5%	95.0%	55	$23,000
0.5%	95.0%	56	$23,700
0.5%	95.0%	57	$24,500
0.5%	95.0%	58	$25,200
0.5%	95.0%	59	$26,000
0.5%	95.0%	60	$26,800
0.5%	95.0%	61	$27,500
0.5%	95.0%	62	$28,300
0.5%	95.0%	63	$29,000
0.5%	95.0%	64	$29,800
0.5%	95.0%	65	$30,500
0.5%	95.0%	66	$31,300
0.5%	95.0%	67	$32,000
0.5%	95.0%	68	$32,800
0.5%	95.0%	69	$33,500
0.5%	95.0%	70	$34,300
0.5%	95.0%	71	$35,000
0.5%	95.0%	72	$35,800
0.5%	95.0%	73	$36,500
0.5%	95.0%	74	$37,300
0.5%	95.0%	75	$38,000
0.5%	95.0%	76	$38,700
0.5%	95.0%	77	$39,500
0.5%	95.0%	78	$40,200
0.5%	95.0%	79	$41,000
0.5%	95.0%	80	$41,700
0.5%	95.0%	81	$42,400
0.5%	95.0%	82	$43,100
0.5%	95.0%	83	$43,800
0.5%	95.0%	84	$44,600
0.5%	95.0%	85	$45,300
0.5%	95.0%	86	$46,000
0.5%	95.0%	87	$46,600
0.5%	95.0%	88	$47,300
0.5%	95.0%	89	$48,000
0.5%	95.0%	90	$48,700
0.5%	95.0%	91	$49,400
0.5%	95.0%	92	$50,000
0.5%	95.0%	93	$50,700
0.5%	95.0%	94	$51,400
0.5%	95.0%	95	$52,000
0.5%	95.0%	96	$52,700
0.5%	95.0%	97	$53,400

Table A.3.8. Estimated lower tolerance bound yield strength as function of Rockwell B hardness, confidence level, and percentile — based on collected data with tensile and hardness property measures for transmission line joints manufactured prior to 1980 (Continued).

Targeted Percentile	Confidence Level	Measured Hardness (HRB)	Estimated Lower Tolerance Bound on Yield Strength (PSI)
1.0%	99.9%	54	21,500
1.0%	99.9%	55	22,300
1.0%	99.9%	56	23,100
1.0%	99.9%	57	23,900
1.0%	99.9%	58	24,700
1.0%	99.9%	59	25,500
1.0%	99.9%	60	26,300
1.0%	99.9%	61	27,100
1.0%	99.9%	62	27,900
1.0%	99.9%	63	28,700
1.0%	99.9%	64	29,500
1.0%	99.9%	65	30,400
1.0%	99.9%	66	31,200
1.0%	99.9%	67	32,000
1.0%	99.9%	68	32,800
1.0%	99.9%	69	33,600
1.0%	99.9%	70	34,400
1.0%	99.9%	71	35,200
1.0%	99.9%	72	35,900
1.0%	99.9%	73	36,700
1.0%	99.9%	74	37,500
1.0%	99.9%	75	38,300
1.0%	99.9%	76	39,100
1.0%	99.9%	77	39,900
1.0%	99.9%	78	40,600
1.0%	99.9%	79	41,400
1.0%	99.9%	80	42,200
1.0%	99.9%	81	42,900
1.0%	99.9%	82	43,700
1.0%	99.9%	83	44,400
1.0%	99.9%	84	45,100
1.0%	99.9%	85	45,800
1.0%	99.9%	86	46,500
1.0%	99.9%	87	47,200
1.0%	99.9%	88	47,900
1.0%	99.9%	89	48,600
1.0%	99.9%	90	49,300
1.0%	99.9%	91	49,900
1.0%	99.9%	92	50,500
1.0%	99.9%	93	51,100
1.0%	99.9%	94	51,800
1.0%	99.9%	95	52,400
1.0%	99.9%	96	53,000
1.0%	99.9%	97	53,600

Table A.3.8. Estimated lower tolerance bound yield strength as function of Rockwell B hardness, confidence level, and percentile — based on collected data with tensile and hardness property measures for transmission line joints manufactured prior to 1980 (Continued).

Targeted Percentile	Confidence Level	Measured Hardness (HRB)	Estimated Lower Tolerance Bound on Yield Strength (PSI)
1.0%	99.0%	54	22,500
1.0%	99.0%	55	23,300
1.0%	99.0%	56	24,000
1.0%	99.0%	57	24,800
1.0%	99.0%	58	25,600
1.0%	99.0%	59	26,400
1.0%	99.0%	60	27,200
1.0%	99.0%	61	27,900
1.0%	99.0%	62	28,700
1.0%	99.0%	63	29,500
1.0%	99.0%	64	30,300
1.0%	99.0%	65	31,100
1.0%	99.0%	66	31,800
1.0%	99.0%	67	32,600
1.0%	99.0%	68	33,400
1.0%	99.0%	69	34,200
1.0%	99.0%	70	34,900
1.0%	99.0%	71	35,700
1.0%	99.0%	72	36,500
1.0%	99.0%	73	37,200
1.0%	99.0%	74	38,000
1.0%	99.0%	75	38,800
1.0%	99.0%	76	39,600
1.0%	99.0%	77	40,300
1.0%	99.0%	78	41,000
1.0%	99.0%	79	41,800
1.0%	99.0%	80	42,500
1.0%	99.0%	81	43,300
1.0%	99.0%	82	44,000
1.0%	99.0%	83	44,700
1.0%	99.0%	84	45,500
1.0%	99.0%	85	46,200
1.0%	99.0%	86	46,900
1.0%	99.0%	87	47,600
1.0%	99.0%	88	48,200
1.0%	99.0%	89	48,900
1.0%	99.0%	90	49,600
1.0%	99.0%	91	50,200
1.0%	99.0%	92	50,900
1.0%	99.0%	93	51,500
1.0%	99.0%	94	52,200
1.0%	99.0%	95	52,800
1.0%	99.0%	96	53,500
1.0%	99.0%	97	54,100

Table A.3.8. **Estimated lower tolerance bound yield strength as function of Rockwell B hardness, confidence level, and percentile — based on collected data with tensile and hardness property measures for transmission line joints manufactured prior to 1980 (Continued).**

Targeted Percentile	Confidence Level	Measured Hardness (HRB)	Estimated Lower Tolerance Bound on Yield Strength (PSI)
1.0%	95.0%	54	23,400
1.0%	95.0%	55	24,100
1.0%	95.0%	56	24,900
1.0%	95.0%	57	25,600
1.0%	95.0%	58	26,400
1.0%	95.0%	59	27,100
1.0%	95.0%	60	27,900
1.0%	95.0%	61	28,700
1.0%	95.0%	62	29,400
1.0%	95.0%	63	30,200
1.0%	95.0%	64	30,900
1.0%	95.0%	65	31,700
1.0%	95.0%	66	32,400
1.0%	95.0%	67	33,200
1.0%	95.0%	68	33,900
1.0%	95.0%	69	34,700
1.0%	95.0%	70	35,400
1.0%	95.0%	71	36,200
1.0%	95.0%	72	36,900
1.0%	95.0%	73	37,700
1.0%	95.0%	74	38,400
1.0%	95.0%	75	39,200
1.0%	95.0%	76	40,000
1.0%	95.0%	77	40,700
1.0%	95.0%	78	41,400
1.0%	95.0%	79	42,100
1.0%	95.0%	80	42,900
1.0%	95.0%	81	43,600
1.0%	95.0%	82	44,300
1.0%	95.0%	83	45,000
1.0%	95.0%	84	45,700
1.0%	95.0%	85	46,400
1.0%	95.0%	86	47,100
1.0%	95.0%	87	47,800
1.0%	95.0%	88 .	48,500
1.0%	95.0%	89	49,200
1.0%	95.0%	90	49,900
1.0%	95.0%	91	50,500
1.0%	95.0%	92	51,200
1.0%	95.0%	93	51,900
1.0%	95.0%	94	52,500
1.0%	95.0%	95	53,200
1.0%	95.0%	96	53,900
1.0%	95.0%	97	54,500